Lecture Notes in Control and Information Sciences

Edited by M. Thoma and A. Wyner

For information about Vols. 1–61 please contact your bookseller or Springer-Verlag.

Lecture Notes in Control and Information Sciences

Edited by M. Thoma and A. Wyner

117

K.J. Hunt

Stochastic Optimal Control Theory
with Application
in Self-Tuning Control

Springer-Verlag Berlin Heidelberg GmbH

Series Editors

M. Thoma · A. Wyner

Advisory Board

L. D. Davisson · A. G. J. MacFarlane · H. Kwakernaak
J. L. Massey · Ya Z. Tsypkin · A. J. Viterbi

Author

Kenneth J. Hunt
BBN Systems and Technologies
Heriot-Watt Research Park
Riccarton, Edinburgh EH 14 4AP
Scotland

ISBN 978-3-540-50532-7   ISBN 978-3-540-46040-4 (eBook)
DOI 10.1007/978-3-540-46040-4
Library of Congress Cataloging in Publication Data

Hunt, K. J. (Kenneth J.)
Stochastic optimal control theory with application in self-tuning control / K.J. Hunt.
(Lecture notes in control and information sciences ; 117)
Bibliography: p.

1. Self tuning controllers. 2. Control theory. 3. Stochastic processes. I. Title. II. Series.
TJ217.H86 1989
629.8'312—dc19   88-37329

Originally published by Springer-Verlag Berlin Heidelberg New York in 1989.

2161/3020-543210

PREFACE

The design of control systems by the mathematical optimisation
of a specified quantitative performance criterion has a long and well
established role in the history of control engineering. The most
widely studied class of optimal control laws are those based upon the
state-space system model. An alternative approach which has been
developed more recently is the design of optimal controllers using
the algebra of polynomials and polynomial matrices. In this
approach scalar systems are described in transfer-function form using
ratios of polynomials, while multivariable systems are described
using left and right matrix factorisations.

A further major branch of control engineering, adaptive control,
has grown from the need to design control systems subject to the
practical constraint of plants whose dynamics are uncertain or
time-varying. Of the various classes of adaptive controllers which
exist, self-tuning control has emerged as perhaps the most widely
studied and applied.

This book merges the above two major areas of control :
original contributions are made in the polynomial approach to
stochastic optimal control theory (LQG control) and the resulting
control laws are then manipulated into a form suitable for
application in the self-tuning control framework. The results of an
application study in which the LQG self-tuner was tested on the steam
pressure control loop of a power station are presented.

Layout

    The work is divided into four parts which are made up of a total of six chapters. Each part concentrates on a different aspect of the overall theme. The parts are organised in such a way that the book follows a logical development from theoretical derivation through to self-tuning controller application:

    Part 1 : Stochastic Optimal Control Theory

    Part 2 : Self-tuning Control

    Part 3 : Case Study

    Part 4 : Conclusions

Part 1, Stochastic Optimal Control Theory, consists of Chapters 1 and 2 and is mainly theoretical in nature. Chapter 1 is an introduction to Part 1 while Chapter 2 develops some original theoretical results in optimal control theory and in particular the derivation of controllers for the optimal rejection of measurable disturbances using feedforward. One major original contribution of Chapter 2 is the extension of existing results to the case of unstable reference and disturbance generating sub-systems. This extension is of major practical importance since it is the unstable generators used to model step-like and deterministic signals which arise in most applications. A further important original contribution in Chapter 2 is the derivation of necessary and sufficient conditions for the optimality of the various polynomial equations arising in the optimal controller designs.   Finally, Chapter 2 concludes with the derivation of the optimal feedback/feedforward regulator for multivariable systems.

    Part 2 of the book, Self-tuning Control, consists of Chapters 3 and 4. Chapter 3 is an introduction to Part 2 while in Chapter 4 the theory derived in Chapter 2 is reduced to a practical self-tuning control algorithm. Self-tuning controllers based on polynomial LQG control have previously been considered. The self-tuning algorithm presented in Chapter 4, however, has several novel features : optimal feedforward compensation of measurable disturbances, dynamic cost-function weights, and a three-level design algorithm with a range of complexity. In addition, the precise role of the various polynomial equations arising in the design is clarified using the results derived in Chapter 2.

Part 3 of the book, consisting of Chapter 5, is a case study. The results of an experimental application of LQG self-tuning control on the steam pressure control loop of the Hunterston 'B' power station simulator are presented. The LQG controller is shown to be very simple to commission and to give improved performance over the existing analogue PI controller.

Finally, the book is concluded in Part 4 (Chapter 6).

Acknowledgements

This book was written while the author was a member of the Industrial Control Unit at the University of Strathclyde, Glasgow. The work grew out of an industrially sponsored project concerned with the development of practical LQG self-tuning algorithms. The support of the funding bodies, the South of Scotland Electricity Board and the Scottish Development Agency, is gratefully acknowledged.

CONTENTS

PART 2: Self-Tuning Control

Chapter 3: Introduction to Self-Tuning Control

Summary

PART 4: Conclusions 241

Chapter 6: Conclusions 243

References 247

Appendices 267

NOTATION

All systems considered are assumed to be linear, time-invariant
and discrete-time. The systems are described in the time-domain by
means of real polynomials in the delay operator d, and in the
frequency-domain by means of real polynomials in the inverse of the
z-transform complex number z. A polynomial X(d) is <u>stable</u> (or
<u>strictly Hurwitz</u>) iff it has no zeros with magnitude less than or
equal to unity. A polynomial X(d) is <u>Hurwitz</u> iff it has no zeros
with magnitude less than unity. A polynomial X(d) is <u>unstable</u> iff it
has any zeros with magnitude less than or equal to unity.

For simplicity the arguments of polynomials are often omitted so
that X(d) is denoted by X. The conjugate of a polynomial X(d) is
denoted by $X^*(d) \triangleq X(d^{-1})$, or simply X^*. The absolute coefficient
of X is denoted by $\langle X \rangle$.

A transfer-function is <u>inverse stable</u> ('minimum phase') iff it
has no zeros with magnitude less than or equal to unity.

The power spectrum of a signal x(t) is denoted by ϕ_x.

In the multivariable case described in Section 2.11 the system
is described by means of real polynomial matrices in d. The adjoint
of a polynomial matrix X(d) is denoted by $X^*(d) \triangleq X^T(d^{-1})$. $\langle X \rangle$
denotes the matrix whose elements are the absolute coefficients of
the polynomials in X.

PART ONE

STOCHASTIC OPTIMAL CONTROL THEORY

CHAPTER ONE

INTRODUCTION TO STOCHASTIC OPTIMAL CONTROL

Summary

This chapter provides an introduction to Part 1 of the book. A brief historical review of feedback control and control theory are first given in Sections 1.1 and 1.2, respectively. The movement away from frequency-response methods towards optimisation techniques, which occurred during the fifties, is discussed in Section 1.3. The factors which then led to the predominance of state-space methods during the sixties are described in Section 1.4. A renewed interest in the frequency-response (transfer-function) approach to controller design occurred during the seventies. This trend included Kučera's pioneering work on the polynomial equation approach to stochastic optimal control, and is discussed in Section 1.5. Finally, the contributions made in Chapter 2 in the polynomial approach to optimal control are outlined in Section 1.6.

1.1   THE ORIGINS OF FEEDBACK CONTROL

Feedback is a fundamental biological mechanism which prevails in all interactions between living organisms and their environment. Moreover, the conscious employment of feedback control by humans has a very well established place in the history of humankind. Perhaps the first use of feedback control is recorded in the Encyclopedia Britannica and comes from the Babylonian era of around 4000 years ago (Gadd, 1929). The agricultural production which helped sustain the Babylonians was supported by a sophisticated system of irrigation in which the moisture content of the soil was regulated to a desired level by the manual opening and closing of water ditches.

Although <u>automatic</u> control devices were used earlier (see Usher 1954) the inception of automatic feedback control as a science is widely regarded as occurring during the latter half of the eighteenth century with the arrival of the Industrial Revolution (MacFarlane, 1979). One of the first applications was Meikle's invention of an automatic turning gear for windmills in 1750 (see Wolf, 1938). In order to turn the main sails into the wind an auxiliary windmill at right angles was employed. Any error in the heading of the turret was thereby translated into a mechanical motion which turned the main sails until they received the full wind.

The most celebrated of the early applications of automatic feedback control is Watt's use in 1788 of the flyball governor for regulating the speed of the steam engine. This device used the principle of the centrifugal governor : a drop in engine speed causes a decrease in the centrifugal force exerted by the flyballs on a spring mechanism and the flyball assembly descends. By lever action this results in the opening of the steam valve which increases the power, and thereby restores the speed, of the engine.

It is clear, therefore, that feedback control systems were used to solve important technical problems long before formal analysis and design techniques appeared. This general lack of a theoretical foundation became apparent in the early nineteenth century as the use of Watt's governor became widespread and as demands for improved performance grew stronger. The increasing tendency for such systems to 'hunt' became apparent; the engine speed displayed a cyclic time variation. The fluctuation of engine speeds remained a major problem for some time. The problem was finally solved in the classic paper by Maxwell (1868) who related system stability to the engineering design parameters. In the steam engine the tendency towards the use of smaller flywheels and increased mass of flyball weights were found to have a destabilising effect.

The early history of feedback control is described by Mayr (1970) and Bennett (1979).

1.2 EARLY DEVELOPMENTS IN CONTROL THEORY

The paper by Maxwell (1868) is regarded as the foundation of the theory of automatic feedback control. Following Maxwell's work the stability problem was treated in terms of differential equations. Routh (1877) and Hurwitz (1895) developed tests to determine the stability of the roots of the system characteristic equation. These tests, however, could only determine absolute system stability and gave no indication of relative stability. The importance of differential equations and their related characteristic equations in control system analysis and design was nevertheless consolidated in the early twentieth century, particularly by the works of Minorsky (1922) on the automatic steering of ships, and Hazen (1934) on servomechanisms.

The most influential work in the development of frequency response methods in control systems was undoubtedly the classic paper by Nyquist (1932). Nyquist's work was again motivated by an important technical problem, that of feedback amplifier stability in long-distance telephony. The implications of Nyquist's Stability

Criterion were, however, much broader than this application. The
frequency response method allowed the gain of feedback control
systems to be set in accordance with specifications on the <u>degree</u> of
stability. Nyquist's criterion was quickly adopted as the main
design tool of control engineers and replaced the earlier time domain
methods based upon differential equations. The trend towards
frequency-response methods was further accelerated by the important
work of Bode (1940) which introduced the concepts of gain and phase
margin. The urgent requirements imposed by the second world war
finally led to the widespread application of these methods when an
immense effort and channelling of resources was directed towards the
development of high performance control systems. A comprehensive
account of the design methods developed at this time is given by
Chestnut and Meyer (1951).

1.3 ANALYTICAL DESIGN METHODS

The design of control systems using the frequency response
methods required a trial-and-error approach whereby the design
procedure was iterated until the performance and stability
specifications were met. During the second world war, however, the
demands for high-precision control led to the first developments in
optimal control theory. The design of servomechanisms by
minimisation of the mean-square tracking error was considered by Hall
(1943) and James et al (1947). However, a comprehensive treatment of
the optimal control problem did not appear until after the work of
Wiener (1949). Wiener had investigated the optimisation of radar

tracking systems where the disturbances were characterised as stochastic processes.

Following the work of Wiener (whose solutions were based upon the so-called Wiener-Hopf integral equation) Newton et al (1957) and Chang (1961) derived optimal controllers based upon the minimisation of integral-type criteria. The term 'analytical design' is defined by Newton et al as 'the design of control systems by application of the methods of mathematical analysis to idealised models which represent physical equipment'. In the analytical designs the system performance is measured by a specified performance index (cost-function) which is normally a weighted sum of error and control input energies. The optimal controller which minimises the performance index is obtained by a systematic procedure of solving the design equations which have been obtained by prior analysis. The analytical design techniques provide a sharp contrast to the trial-and-error methods since they (ideally) proceed from the problem specification directly to the final controller design without the need for subjective analysis. Newton's solution of the optimal control problem using the Wiener-Hopf approach had one major drawback. In the orginal design procedure the equivalent cascade compensator is first found and is then used to calculate the corresponding controller for the feedback loop. This approach, which is inherently open-loop, can yield an unstable closed-loop system unless the controlled process is stable. This is due to pole-zero cancellations within the feedback loop. The general confusion in transfer-function methods surrounding the pole-zero cancellation problem was not resolved until Kučera (1974,1975) and, independently,

Youla et al (1976 a,b) derived a parameterisation for the class of all controllers resulting in a stable closed-loop system. The first full treatment of the Wiener-Hopf optimal controller design for possibly unstable plants was subsequently given by Youla et al (1976a) and generalised to the multivariable case by Youla et al (1976b).

1.4 STATE SPACE OPTIMAL CONTROL

The long gap between the original work of Newton and the proper general solution of the Wiener-Hopf approach to the optimal control problem given by Youla can be attributed to the emergence in the late fifties of state-space methods. These methods employ the mathematical tools of differential equations and vector spaces and admit the exact characterisation of the internal properties and structure of the system (Zadeh and Desoer 1963, Kalman 1963). The maximum principle of Pontryagin (1963) and Bellman's (1957) work on dynamic programming laid the foundations for the treatment of the linear optimal control problem in the state-space (Kalman, 1960). The combination of the new optimal control result with the innovations made by Kalman and Bucy (1961) in filtering theory then led to the celebrated LQG (linear-quadratic-gaussian) design method.

The widespread adoption of the LQG method was established throughout the sixties by work on another major technical problem. In both the USA and USSR a major research and development effort was directed towards the control of space vehicles. During this period LQG optimal control theory became an established design tool for linear systems and several standard texts soon appeared (Athans and

Falb 1966, Bryson and Ho 1969, Anderson and Moore 1971, Kwakernaak and Sivan 1972). The success of the LQG method in the sixties can be attributed to several factors. Firstly, the state-space model employed was immediately applicable to the multivariable situation. Secondly, the nature of the space vehicles being controlled meant that accurate models and measurements were available. Finally, the quadratic form of performance index was often closely correlated with the 'economic' nature of the demanded system performance (such as fuel minimisation).

1.5 THE POLYNOMIAL EQUATION APPROACH

The success of the LQG design method in the aerospace problems of the sixties was not repeated when the techniques were applied to industrial process control problems. The above conditions which contributed to the earlier successes do not in general hold in such situations. Many process control problems are characterised by a high degree of uncertainty in the model available. In addition, the implicit assumption that all state-variables are available for measurement is no longer valid and the need for state reconstruction is hampered by the difficulty of measurement.

These factors led in the seventies to a renewed interest in the frequency-domain (transfer-function) approach to controller design. Some important works in this respect are those of Rosenbrock (1969,1970), Mayne (1973), Wolovich (1974) and Postlethwaite and MacFarlane (1979). The growing presence of algebraic and geometric concepts in system theory was also apparent through the works of

Kalman et al (1969), Wonham (1974) and Bengtsson (1973,1977). As mentioned above the frequency-domain approach to optimal control was generalised at this time by Youla and co-workers whose solutions, however, required rather complicated numerical procedures.

The polynomial equation approach to optimal control design is a transfer-function method which provides an alternative to the Wiener-Hopf technique. The first steps in the polynomial equation design procedure were taken by Åström (1970) and Peterka (1972) with the derivation of a minimum output variance regulator. A comprehensive treatment of the stochastic optimal multivariable control problem using the polynomial equation approach was given in a series of papers throughout the seventies by Kučera, whose pioneering work on the subject culminated in the publication of a book (Kučera 1979). In this approach synthesis of the optimal control law reduces to the solution of linear polynomial equations whose coefficients are obtained by spectral factorisation. Simple computational algorithms for these operations are given by Kučera (1979), Ježek (1982) and Ježek and Kučera (1985).

1.6 CONTRIBUTIONS OF THE PRESENT WORK

The results presented in Chapter 2 generalise Kučera's polynomial equation solution of the stochastic optimal control problem. Kučera (1979) addressed the multivariable <u>regulator</u> problem using polynomial techniques. A theory was later developed for both the deterministic and stochastic <u>tracking</u> problems for scalar (single-input, single-output) systems (Šebek 1982, Kučera and Šebek 1984a,b). Šebek, meanwhile, had derived a solution for the multivariable stochastic tracking problem (Šebek 1983a,b).

Grimble (1986a,b) has made several contributions in this area. Shaked (1976), Gupta (1980) and Anderson et al (1983,1985) had previously introduced the concept of dynamic (frequency-dependent) cost-function weighting elements for the state-space LQG design and Grimble incorporated this idea into his generalisation of Kučera's work. A further innovation made by Grimble was the introduction of a coloured output disturbance signal (measurement noise). A major limitation of Grimble's work, however, was the restriction of all disturbance and reference sub-systems to be asymptotically stable (the unstable systems which model signals such as steps, ramps, sinusoids and deterministic signals are of greatest practical importance).

The most significant of Grimble's contributions was the incorporation into the overall design procedure of a feedforward compensator for the rejection of <u>measurable</u> disturbances (Grimble 1986b). This analysis was again limited to the case of asymptotically stable disturbance and reference generating sub-systems. The general solution of the feedforward problem was subsequently given by Šebek, Hunt and Grimble (1988) for the case of scalar cost-function weights and white measurement noise. A polynomial solution to the feedforward problem has been independently obtained by Sternad (1985,1987) using an alternative proof technique. Sternad's analysis is for the case of scalar cost weights, zero measurement noise and an asymptotically stable measurable disturbance generator.

The first contribution of the present work is that the basic system model considered extends the results obtained by Grimble to the case of possibly unstable reference and disturbance sub-systems. In addition, the optimal control problem is solved for both the single-degree-of-freedom and two-degrees-of-freedom controller structures. To summarise, the problem considered is as follows:

(i)   The cost-function includes dynamic weighting elements.

(ii)   The system model includes a coloured output disturbance signal (measurement noise).

(iii) A feedforward compensator is incorporated in the overall design procedure for the rejection of measurable disturbances.

(iv)   All disturbance and generating sub-systems may be unstable.

(v)   Solutions are obtained for both the single and two-degrees-of-freedom controller structures (including, in each case, feedforward).

The extension to the case of unstable disturbance and reference sub-systems is non-trivial since this involves the derivation of appropriate necessary and sufficient problem solvability conditions (the restriction to stable sub-systems is, in fact, sufficient to ensure problem solvability).

The optimal controller results presented in Chapter 2 for the single-degree-of-freedom case are summarised in Hunt (1988a) and for the two-degree-of-freedom case in Hunt (1988b).

The general solution of the optimal control problem for both the single- and two-degrees-of-freedom control structures requires that a couple of polynomial equations be solved to obtain each part of the controller (in the two-degrees-of-freedom structure the controller consists of three parts: a reference part, a feedback part and a feedforward part). By eliminating the common term between each couple of equations a single, related, equation is obtained (the so-called 'implied' equations). The second major contribution of Chapter 2 is the derivation of the conditions under which the implied polynomial equations may be solved to obtain the unique optimal controller polynomials. The conditions relating to optimality of the implied feedback and reference equations in the case of scalar cost-function weights have been known for some time (Šebek and Kučera 1982, Kučera 1984). The corresponding result for the feedback equation in the multivariable case (with dynamic weights) has recently been derived by Hunt, Šebek and Grimble (1987). The derivation of the conditions relating to optimality of the implied feedback and reference equations in the case of dynamic weights given in Chapter 2 extends the previous results. The analysis for the implied feedforward equation is completely new. Roberts (1986, 1987 a,b,c,d,e) has investigated a related problem regarding the sufficiency of the first equation in the couple of feedback equations.

These results are followed by a summary of the important structural properties of the optimal control designs.

Finally, the optimal feedback/feedforward regulator solution of Šebek, Hunt and Grimble (1988) is extended to the multivariable case. This new multivariable result is also summarised in Hunt and Šebek (1989).

A final word on the use of the phrase 'Stochastic Optimal Control' in place of 'LQG Control' in the title of this work : in the state-space LQG methods the restriction of the noise sources to be Gaussian distributed is required in order that the optimal control law, which is chosen from the set of all (possibly non-linear) controllers, is a <u>linear</u> state feedback. In the polynomial equation approach (as in the Wiener-Hopf method) the controller is assumed at the outset to be linear. The gaussian restriction, therefore, is no longer required (this argument is due to Kučera, 1987). However, it has become standard practice in the polynomial equation approach to use 'LQG control' synonymously with 'Stochastic Optimal Control' and this convention will be adopted throughout the remainder of this work.

CHAPTER TWO

STOCHASTIC TRACKING WITH MEASURABLE DISTURBANCE FEEDFORWARD

Summary

The open-loop model for the single-input single-output plant under consideration is described in Section 2.1. The plant output which is to be controlled is affected by two disturbance signals, one of which is assumed measurable. Associated with the measurable disturbance is a white measurement noise. Associated with the measurement of the controlled output is a measurement noise, or output disturbance, which may be coloured. For tracking purposes a reference, or command, signal is introduced. This signal is again corrupted by a measurement noise.

Two types of controller structure are introduced in Section 2.2:

(i) The single-degree of freedom (SDF) structure where the observed tracking error is processed by a single controller in cascade with the plant.

(ii) The two-degrees-of-freedom (2DF) structure where the measured reference and measured output signals are processed independently by a reference and feedback controller, respectively.

In both the SDF and 2DF structures a feedforward compensator is also employed to counter the effect of the measurable disturbance.

In Section 2.3 the optimal control problem is defined by specifying the cost-function which is to be minimised. A feature of the cost-function employed is the inclusion of dynamic (frequency-dependent) weighting elements. The general problem for the SDF and 2DF control structures including feedforward is solved in Sections 2.4 and 2.5, respectively. Also included is an analysis of the problem of internal stability for the resulting closed-loop systems. In Sections 2.6 and 2.7 the general problem for the SDF and 2DF structures is re-solved for the case when the plant is expressed using a least-common-denominator polynomial for each of its sub-systems. While the original solution provides insight into the role played by each individual sub-system the common denominator solution is computationally more efficient. The two solutions are, of course, exactly equivalent. The complete general solution of the optimal control problem requires that a couple of polynomial equations be solved to obtain each part of the controller. By eliminating the common term between each couple of equations a single, related, equation is obtained (the so called 'implied' equations). The conditions under which the implied polynomial equations may be solved to obtain the unique optimal controller polynomials are derived for the SDF and 2DF structures in Sections 2.8 and 2.9, respectively. A summary of the main structural properties of the optimal controller solutions is given in Section 2.10.

The chapter concludes in Section 2.11 with the derivation of the optimal feedback/feedforward regulator for <u>multivariable</u> plants.

2.1   PLANT MODEL

The open-loop model for the single-input single-output <u>plant</u> under consideration is shown in Figure 2.1. The plant is governed by the equations:

$$y(t) = p(t) + x(t) + d(t) \qquad\qquad (2.1)$$

$$= W_p u(t) + W_x \ell(t) + W_d \psi_d(t) \qquad\qquad (2.2)$$

The controlled output, $y(t)$, consists of the sum of three signals:

(i)   The 'undisturbed' output $p(t) = W_p u(t)$, where $u(t)$ is the plant <u>control input</u>.

(ii)   A disturbance signal $x(t) = W_x \ell(t)$, where $\ell(t)$ is a <u>measurable disturbance</u>.

(iii)   A disturbance signal $d(t) = W_d \psi_d(t)$, where ψ_d is an <u>unmeasurable</u> stochastic signal.

The controlled output is corrupted by a <u>measurement noise</u> $n(t)$. The measured output, $z(t)$, is given by the equations:

$$z(t) = y(t) + n(t) \qquad\qquad (2.3)$$

$$= y(t) + W_n \psi_n(t) \qquad\qquad (2.4)$$

where $\psi_n(t)$ is an <u>unmeasurable</u> stochastic signal.

The measurable disturbance signal $\ell(t)$ is corrupted by a stochastic <u>measurement noise</u> $\psi_{\ell n}(t)$. The <u>disturbance measurement</u>, $f(t)$, is given by:

$$f(t) = \ell(t) + \psi_{\ell n}(t) \qquad\qquad (2.5)$$

The open-loop plant structure shown in Figure 2.1 is representative of many industrial control problems:

(i)   The signal $\ell(t)$ typically represents a load disturbance which can be measured and used to provide feedforward control. The signal $\psi_{\ell n}(t)$ represents noise arising from the measurement of $\ell(t)$, so that the actual signal used for feedforward is $f(t)$.

(ii) The measured output available for feedback ($z(t)$) is usually different from the output to be controlled ($y(t)$) due to measurement noise which the controller should not attempt to regulate. Use of the filter W_n admits the modelling of many different forms of measurement noise. For example, in ship control systems $n(t)$ represents the high-frequency effect of waves to which the controller should not respond (see Grimble, 1986a).

Polynomial form

The transfer-functions of the various sub-systems in the plant model may be represented as ratios of polynomials in the delay operator d as follows:

$$W_p = A_p^{-1} B_p \qquad\qquad (2.6)$$

$$W_d = A_d^{-1} C_d \qquad\qquad (2.7)$$

$$W_x = A_x^{-1} C_x \qquad\qquad (2.8)$$

$$W_n = A_n^{-1} C_n \qquad\qquad (2.9)$$

Any common factors of A_d and A_x are denoted by D_{dx} such that:

$$A_d = A_d' D_{dx} \; , \; A_x = A_x' D_{dx} \qquad\qquad (2.10)$$

The least common multiple of A_d and A_x is denoted by A_{dx} i.e:

$$l.c.m(A_d, A_x) = A_{dx} = A_d' A_x' D_{dx} \qquad\qquad (2.11)$$

Any common factors of A_{dx} and A_p are denoted by D_{pdx} such that:

$$A_{dx} = A_{dx}' D_{pdx} \; , \; A_p = A_p' D_{pdx} \qquad\qquad (2.12)$$

Command signal model

In the optimal tracking control problem considered in the following the controlled output y(t) will be required to follow as closely as possible a _reference_ (or command) signal r(t). The signal r(t) may be represented as the output of a _generating sub-system_ W_r which is driven by an external stochastic signal $\psi_r(t)$:

$$r(t) = W_r \psi_r(t) \tag{2.13}$$

The sub-system W_r is represented in polynomial form as:

$$W_r = A_e^{-1} E_r \tag{2.14}$$

where A_e and E_r are polynomials in d.

The reference signal r(t) is corrupted by a stochastic _measurement noise_ $\psi_{rn}(t)$. The _reference measurement_, m(t), is given by:

$$m(t) = r(t) + \psi_{rn}(t) \tag{2.15}$$

The _tracking error_, e(t), is defined by:

$$e(t) \triangleq r(t) - y(t) \tag{2.16}$$

Any common factors of A_e and A_p are denoted by D_{pe} such that:

$$A_e = A_e' D_{pe} \ , \ A_p = A_{pe}' D_{pe} \tag{2.17}$$

The least common multiple of A_d, A_e and A_x is denoted by A_{dex} i.e.:

$$\ell.c.m \ (A_d, \ A_e, \ A_x) = A_{dex} \tag{2.18}$$

Any common factors of A_{dex} and A_p are denoted by D_{dex} such that:

$$A_{dex} = A_{dexp}' D_{dex} \ , \ A_p = A_{pf}' D_{dex} \tag{2.19}$$

Denote by A_{ex} and A_{ed} the least common multiples of A_e, A_x and A_e, A_d respectively, i.e:

$$A_{ex} = \ell.c.m \ (A_e, A_x) \tag{2.20}$$

$$A_{ed} = \ell.c.m \ (A_e, A_d) \tag{2.21}$$

Any common factors of A_{ex} and A_d are denoted by D_{exd} such that:

$$A_{ex} = A'_{ex}D_{exd}, \quad A_d = A'_{dex}D_{exd} \tag{2.22}$$

Any common factors of A_{dx} and A_e are denoted by D_{xde} such that:

$$A_{dx} = A'_{xd}D_{xde}, \quad A_e = A'_{exd}D_{xde} \tag{2.23}$$

Any common factors of A_{ed} and A_x are denoted by D_{edx} such that:

$$A_{ed} = A'_{ed}D_{edx}, \quad A_x = A'_{xed}D_{edx} \tag{2.24}$$

The reference generator model is shown in Figure 2.2.

Measurable disturbance model

The measurable disturbance signal $\ell(t)$ may be represented as the output of a <u>generating sub-system</u> W_ℓ driven by an external stochastic signal $\psi_\ell(t)$:

$$\ell(t) = W_\ell \psi_\ell(t) \tag{2.25}$$

The sub-system W_ℓ is represented in polynomial form as:

$$W_\ell = A_\ell^{-1} E_\ell \tag{2.26}$$

where A_ℓ and E_ℓ are polynomials in d.

Any common factors of $A_\ell A_x$ and A_p are denoted by $D_{p\ell x}$ so that:

$$A_\ell A_x = A'_{\ell x}D_{p\ell x}, \quad A_p = A'_{p\ell x}D_{p\ell x} \tag{2.27}$$

Assumptions

The following assumptions are made on the various sub-systems defined above:

(i)   Each of the sub-systems is free of unstable hidden modes.

(ii)   The plant input-output transfer-function W_p is assumed strictly causal i.e. $\langle B_p \rangle = 0$.

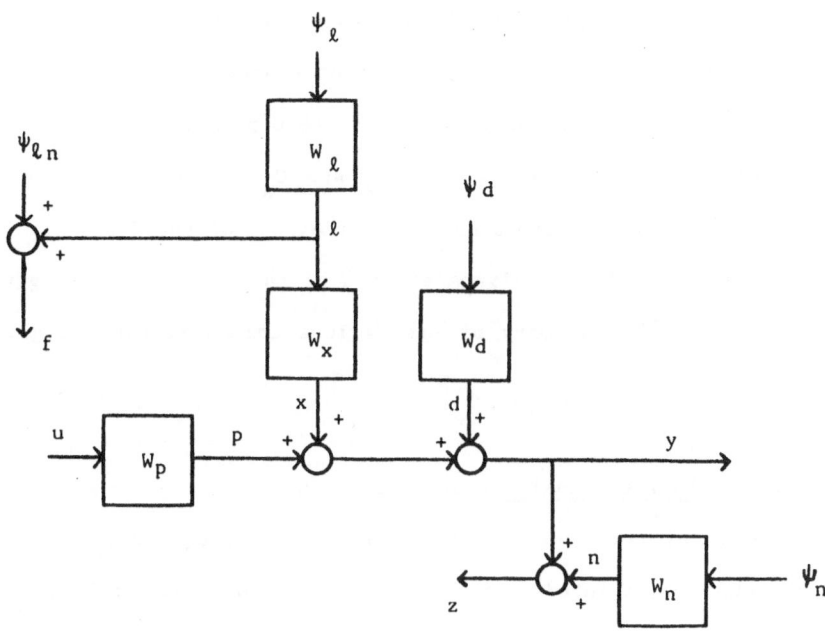

Figure 2.1 : Open-loop plant

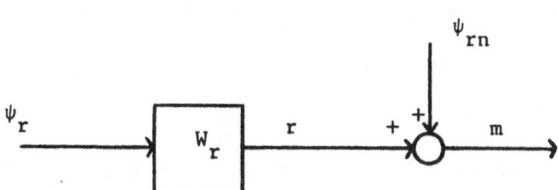

Figure 2.2 : Reference generator

(iii)   The polynomial A_n may have zeros on the unit circle but is assumed, without loss of generality, to have no zeros within the unit circle of the d-plane.

(iv)   The polynomials C_x, C_d and C_n are assumed to have no common factors on the unit circle of the d-plane.

A summary of the polynomial form of each sub-system is given in Table 2.1, and a summary of the various common factors is given in Table 2.2.

2.1.1 Stochastic signal definitions

Each of the stochastic signals ψ_d, ψ_n, ψ_ℓ and ψ_r are assumed mutually uncorrelated and belong to one of the following three categories:

(a)   Stationary white noise signal, where:

(i)   The signal is a sequence of independent, equally distributed random variables i.e. it is a white noise sequence.

(ii)   The signal has zero-mean

(iii)   The signal is wide-sense stationary.

(b)   Non-Stationary signal, where:

(i)   The signal is a compound (or generalised) Poisson process i.e. a sequence consisting of random pulses of magnitude a_i occuring at random times t_i.

(ii)   The random variables a_i form a white noise sequence of the type defined in (a) above.

$$W_p = A_p^{-1} B_p$$

$$W_d = A_d^{-1} C_d$$

$$W_x = A_x^{-1} C_x$$

$$W_n = A_n^{-1} C_n$$

$$W_r = A_e^{-1} E_r$$

$$W_\ell = A_\ell^{-1} E_\ell$$

Table 2.1 : Plant Transfer-functions

	Polynomial pair	Common factor	Factorised pair
	A_d , A_x	D_{dx}	$A_d = A_d' D_{dx}$, $A_x = A_x' D_{dx}$
$A_{dx} = \mathrm{lcm}(A_d, A_x)$	A_{dx} , A_p	D_{pdx}	$A_{dx} = A_{dx}' D_{pdx}$, $A_p = A_p' D_{pdx}$
	A_e , A_p	D_{pe}	$A_e = A_e' D_{pe}$, $A_p = A_{pe}' D_{pe}$
	$A_\ell A_x$, A_p	$D_{p\ell x}$	$A_\ell A_x = A_{\ell x}' D_{p\ell x}$, $A_p = A_{p\ell x}' D_{p\ell x}$
$A_{dex} = \mathrm{lcm}(A_d, A_e, A_x)$	A_{dex} , A_p	D_{dex}	$A_{dex} = A_{dexp}' D_{dex}$, $A_p = A_{pf}' D_{dex}$
$A_{ex} = \mathrm{lcm}(A_e, A_x)$	A_{ex} , A_d	D_{exd}	$A_{ex} = A_{ex}' D_{exd}$, $A_d = A_{dex}' D_{exd}$
	A_{dx} , A_e	D_{xde}	$A_{dx} = A_{xd}' D_{xde}$, $A_e = A_{exd}' D_{xde}$
$A_{ed} = \mathrm{lcm}(A_e, A_d)$	A_{ed} , A_x	D_{edx}	$A_{ed} = A_{ed}' D_{edx}$, $A_x = A_{xed}' D_{edx}$

Table 2.2 : Common Factors

(c) <u>Pulse sequence</u>, where

$$\psi_{.}(t) = \begin{cases} 1 & t = 0 \\ 0 & t \neq 0 \end{cases}$$

The stochastic measurement noise signals $\psi_{\ell n}$ and ψ_{rn} are assumed to be mutually uncorrelated <u>white noise</u> sequences of the type defined in (a) above.

The intensities of the signals ψ_d, ψ_n, ψ_ℓ, ψ_r, $\psi_{\ell n}$ and ψ_{rn} are denoted by σ_d, σ_n, σ_ℓ, σ_r, $\sigma_{\ell n}$ and σ_{rn}, respectively. All intensities are assumed non-zero.

2.1.2 <u>Non-stationary and deterministic reference and disturbances</u>

The above definitions admit the modelling of many different types of reference and disturbance signals $r(t)$, $\ell(t)$, $d(t)$ and $n(t)$. Of considerable practical importance are coloured noise signals, step-like signals, and deterministic signals such as steps, ramps or sinusoids. These types of reference and disturbance signals may be modelled by appropriate definition of the driving sources ψ_r, ψ_ℓ, ψ_d and ψ_n, and of the associated filters W_r, W_ℓ, W_d and W_n, respectively, as follows:

(i)   <u>Coloured zero-mean noise</u> is generated when the driving source is white noise of the type defined in (a) above, and when the filter is asymptotically stable.

(ii)   <u>Random walk</u> sequences are generated when the driving source is white noise of the type defined in (a) above, and when the filter has a denominator $1-d$ (an integrator).

(iii) <u>Step-like</u> sequences consisting of random steps at random times are generated when the driving source is a compound Poisson process of the type defined in (b) above, and when the filter has a denominator 1-d.

(iv) <u>Deterministic</u> sequences such as steps, ramps or sinusoids are generated when the driving source is a unit pulse sequence, and when the filter has poles on the unit circle of the d-plane.

2.2   CONTROL STRUCTURES

Two types of control structure will be considered : the single-degree-of-freedom (SDF) structure and the two-degrees-of-freedom (2DF) structure. In both cases a feedforward compensator will also be employed to counter the effect of the measurable disturbance $\ell(t)$.

In the single-degree-of-freedom structure the <u>observed</u> tracking error is processed by a single cascade compensator, while in the two-degrees-of-freedom structure the observed reference and observed output signals are processed independently. The extra degree of freedom provided by the 2DF structure leads to the following advantages:

(i)   The command response can be shaped independently of the feedback properties of the system.

(ii)   A lower optimal cost can be achieved, (Gawthrop, 1978).

However, in some practical situations it is not possible to realise a 2DF control structure since it is not always possible to measure the reference and output signals separately. For example, in many trajectory following problems it is only possible to measure the tracking error (i.e. the difference between the desired and actual trajectories) and hence a SDF control structure must be used.

2.2.1   Single-degree-of-freedom controller with feedforward

The closed-loop system for the SDF controller including feedforward is shown in Figure 2.3. The <u>observed error</u> signal $e_o(t)$ is defined by:

27

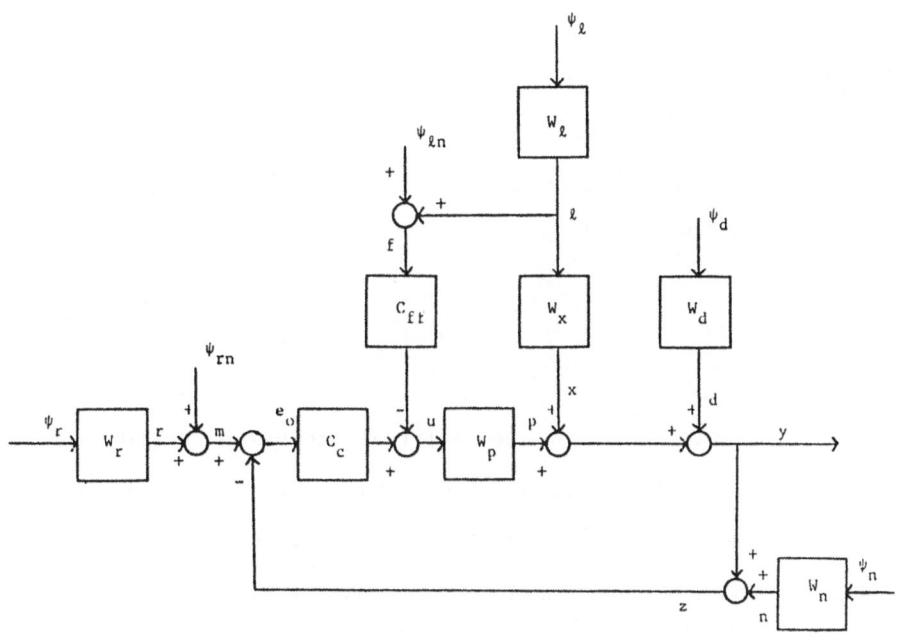

Figure 2.3 : SDF Control System
with Feedforward

$$e_o(t) \triangleq m(t) - z(t) \tag{2.28}$$

The control law is given by:

$$u(t) = C_c e_o(t) - C_{ff} f(t) \tag{2.29}$$

where the <u>cascade controller</u> C_c and the <u>feedforward controller</u> C_{ff} may be expressed as ratios of polynomials in the delay operator d as:

$$C_c = C_{cd}^{-1} C_{cn} \tag{2.30}$$

$$C_{ff} = C_{ffd}^{-1} C_{ffn} \tag{2.31}$$

The transfer-functions from the external stochastic signals to the control input and to the tracking error play a crucial role in determining the solvability of the optimal control problem. For the SDF controller structure shown in Figure 2.3 these transfer-functions are given by:

$$e(t) = - \frac{C_d A_p C_{cd}}{A_d{}^\alpha} \psi_d(t) + \frac{B_p C_{cn} C_n}{A_n{}^\alpha} \psi_n(t)$$

$$+ \frac{A_p C_{cd} E_r}{A_e{}^\alpha} \psi_r(t) - \frac{B_p C_{cn}}{{}^\alpha} \psi_{rn}(t)$$

$$- \frac{(C_x A_p C_{ffd} - B_p C_{ffn} A_x) C_{cd} E_\ell}{A_x A_\ell C_{ffd}{}^\alpha} \psi_\ell(t) + \frac{B_p C_{ffn} C_{cd}}{C_{ffd}{}^\alpha} \psi_{\ell n}(t)$$

$$\tag{2.32}$$

$$u(t) = - \frac{C_d A_p C_{cn}}{A_d{}^\alpha} \psi_d(t) - \frac{A_p C_{cn} C_n}{A_n{}^\alpha} \psi_n(t)$$

$$+ \frac{A_p C_{cn} E_r}{A_e{}^\alpha} \psi_r(t) + \frac{C_{cn} A_p}{{}^\alpha} \psi_{rn}(t)$$

$$- \frac{(C_{cn} C_x C_{ffd} + C_{ffn} A_x C_{cd}) A_p E_\ell}{A_x A_\ell C_{ffd}{}^\alpha} \psi_\ell(t) - \frac{A_p C_{ffn} C_{cd}}{C_{ffd}{}^\alpha} \psi_{\ell n}(t)$$

$$\tag{2.33}$$

where the characteristic polynomial $\propto$ is defined by:

$$\propto \triangleq A_p C_{cd} + B_p C_{cn} \qquad (2.34)$$

2.2.2 Two-degrees-of freedom controller with feedforward

The closed-loop system for the 2DF controller including feedforward is shown in Figure 2.4. The control law is given by:

$$u(t) = - C_{fb} z(t) + C_r m(t) - C_{ff} f(t) \qquad (2.35)$$

where the feedback controller C_{fb}, the reference controller C_r, and the feedforward controller C_{ff} may be expressed as ratios of polynomials in the delay operator d as:

$$C_{fb} = C_{fbd}^{-1} C_{fbn} \qquad (2.36)$$

$$C_r = C_{rd}^{-1} C_{rn} \qquad (2.37)$$

$$C_{ff} = C_{ffd}^{-1} C_{ffn} \qquad (2.38)$$

The transfer-functions from the external stochastic signals to the control input and to the tracking error again play a crucial role in determining the optimal control problem solvability. For the 2DF controller structure shown in Figure 2.4 these transfer-functions are given by:

$$e(t) = - \frac{C_d A_p C_{fbd}}{A_d \propto} \psi_d(t) + \frac{B_p C_{fbn} C_n}{A_n \propto} \psi_n(t)$$

$$+ \frac{(C_{rd} \propto - B_p C_{rn} C_{fbd}) E_r}{A_e \propto C_{rd}} \psi_r(t) - \frac{B_p C_{rn} C_{fbd}}{C_{rd} \propto} \psi_{rn}(t)$$

$$- \frac{(C_x A_p C_{ffd} - A_x C_{ffn} B_p) E_\ell C_{fbd}}{A_x A_\ell C_{ffd} \propto} \psi_\ell(t) + \frac{B_p C_{ffn} C_{fbd}}{C_{ffd} \propto} \psi_{\ell n}(t)$$

$$(2.39)$$

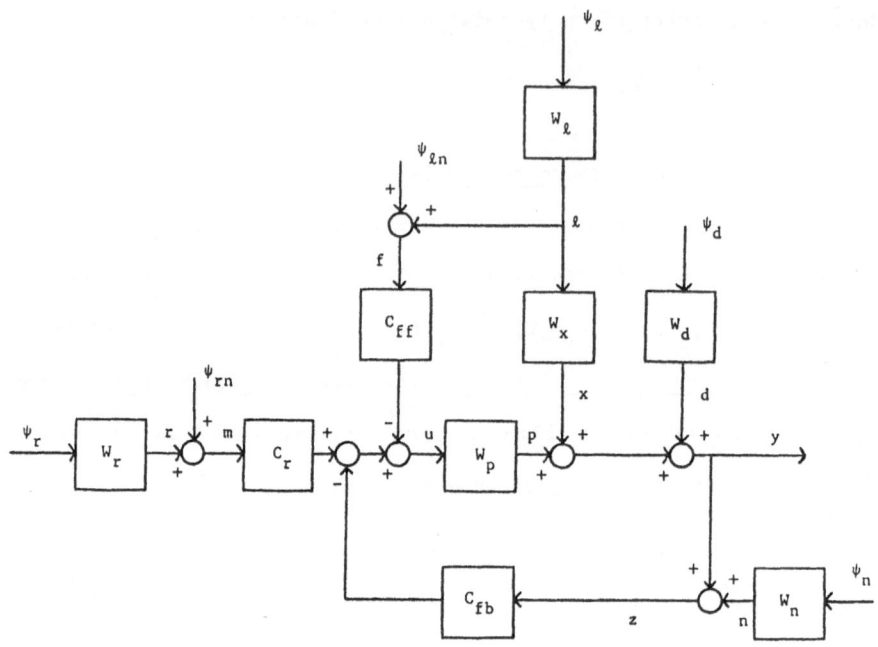

Figure 2.4 : 2DF Control System
with Feedforward

$$u(t) = -\frac{C_{fbn}C_d A_p}{A_d{}^{\alpha}}\,\psi_d(t) - \frac{C_{fbn}C_n A_p}{A_n{}^{\alpha}}\,\psi_n(t)$$

$$+\frac{C_{rn}E_r A_p C_{fbd}}{C_{rd}A_e{}^{\alpha}}\,\psi_r(t) + \frac{C_{rn}A_p C_{fbd}}{C_{rd}{}^{\alpha}}\,\psi_{rn}(t)$$

$$-\frac{(C_{fbn}C_x C_{ffd} + A_x C_{ffn}C_{fbd})E_{\ell}A_p}{A_x A_{\ell}C_{ffd}{}^{\alpha}}\,\psi_{\ell}(t) - \frac{A_p C_{ffn}C_{fbd}}{C_{ffd}{}^{\alpha}}\,\psi_{\ell n}(t)$$

$$(2.40)$$

where the characteristic polynomial α is defined by:

$$\alpha \triangleq A_p C_{fbd} + B_p C_{fbn} \qquad\qquad (2.41)$$

2.3 OPTIMAL CONTROL PROBLEM DEFINITION

The desired optimal SDF and 2DF controllers evolve from minimisation of the cost-function:

$$J = E\{(H_q e)^2(t) + (H_r u)^2(t)\} \tag{2.42}$$

where H_q and H_r are dynamic (i.e. frequency-dependent) weighting elements which may be realised by rational transfer-functions.

Using Parseval's theorem the cost-function may be transformed to the frequency domain and expressed as:

$$J = \frac{1}{2\pi j} \oint_{|z|=1} \{Q_c \phi_e + R_c \phi_u\} \frac{dz}{z} \tag{2.43}$$

where ϕ_e and ϕ_u are the tracking error and control input spectral densities, respectively, and:

$$Q_c = H_q H_q^*, \quad R_c = H_r H_r^* \tag{2.44}$$

The weighting elements Q_c and R_c may be expressed as ratios of polynomials using:

$$Q_c \triangleq \frac{B_q^* B_q}{A_q^* A_q}, \quad R_c \triangleq \frac{B_r^* B_r}{A_r^* A_r} \tag{2.45}$$

Assumptions:

    (i)   The weighting elements Q_c and R_c are strictly positive on $|d|=1$.

    (ii) A_q, B_q, A_r and B_r are strictly Hurwitz polynomials.

2.4 SINGLE-DEGREE-OF-FREEDOM SOLUTION WITH FEEDFORWARD

The <u>Hurwitz</u> spectral factors D_c, D_f and D_{fd} are defined by:

$$D_c D_c^* = B_p A_r B_q B_q^* A_r^* B_p^* + A_p A_q B_r B_r^* A_q^* A_p^* \qquad (2.46)$$

$$D_f D_f^* = (A_n A'_{ex} C_d \sigma_d C_d^* A'^*_{ex} A_n^* + A_{dex} C_n \sigma_n C_n^* A_{dex}^*$$
$$+ A_n A'_{xd} E_r \sigma_r E_r^* A'^*_{xd} A_n^* + A_n A_{dex} \sigma_{rn} A_{dex}^* A_n^*$$
$$+ A_n A'_{ed} C_x \sigma_{\ell n} C_x^* A'^*_{ed} A_n^*) A'_{pf} A'^*_{pf} \qquad (2.47)$$

$$D_{fd} D_{fd}^* = A_\ell \sigma_{\ell n} A_\ell^* + E_\ell \sigma_\ell E_\ell^* \qquad (2.48)$$

<u>Lemma 1</u>:

The polynomials D_c and D_{fd} are <u>strictly Hurwitz</u>

<u>Proof</u>:

Any possible zero of D_c which lies on the unit circle of the d-plane satisfies $d = \exp(j\omega)$. If such a zero exists then, from equation (2.46):

$$B_p A_r B_q(e^{j\omega}) B_p A_r B_q(e^{-j\omega}) + A_p A_q B_r(e^{j\omega}) A_p A_q B_r(e^{-j\omega})$$
$$= |B_p A_r B_q(e^{j\omega})|^2 + |A_p A_q B_r(e^{j\omega})|^2 = 0$$

This implies that $d = \exp(j\omega)$ is a zero of both $B_p A_r B_q$ and $A_p A_q B_r$. The polynomials $B_p A_r B_q$ and $A_p A_q B_r$ cannot, however, have such a zero since A_p and B_p have no unstable common factors (Assumption (1) in Section 2.1) and since A_r, B_q, A_q and B_r are strictly Hurwitz (Assumption (2) in Section 2.3). By contradiction, therefore, D_c is strictly Hurwitz.

By a similar argument any unstable zero of D_{fd} would, from equation (2.48), require:

$$\sigma_{\ell n} |A_\ell(e^{j\omega})|^2 + \sigma_\ell |E_\ell(e^{j\omega})|^2 = 0$$

Since the intensities $\sigma_{\ell n}$ and σ_ℓ are non-zero this condition would require that $d = \exp(j\omega)$ is a zero of both A_ℓ and E_ℓ which again is not possible (Assumption (1) in Section 2.1). As a consequence D_{fd} is strictly Hurwitz.

This proof is based on the proof of Lemma 12.1 in Åström and Wittenmark (1984).

Lemma 2:

Any zeros of A_p which lie on the unit circle and which are not also zeros of A_d, A_x or A_e will be zeros of the spectral factor D_f. If no such zeros exist then D_f is strictly Hurwitz.

Proof:

Any zero of A_p which is not also a zero of A_d, A_x or A_e will, from equation (2.19), be a zero of A'_{pf}. Any such zero lying on the unit circle will, from equation (2.47), also be a zero of D_f.

Using a similar argument to that used in the proof of Lemma 1 the term inside the brackets in equation (2.47) can only lead to strictly Hurwitz terms in D_f.

Theorem 1 : Optimal SDF plus feedforward controller

The optimal control problem for the SDF control structure has a solution if and only if:

(a) Any unstable factors of A_d are also factors of A_p.

(b) Any unstable factors of A_e are also factors of A_p.

(c) Any unstable factors of the product $A_\ell A_x$ are also factors of

A_p.

(d) Any unstable factors of A_n are <u>not</u> also factors of A_p.

The cascade and feedforward parts of the control law (2.29) which

minimises the cost-function (2.43) are as follows:

(i) <u>Optimal cascade controller</u>

$$C_c = \frac{GA_r}{H} \tag{2.49}$$

where G,H (along with F) is the solution having the property:

$$(D_c^* D_f^* z^{-g_1})^{-1} F \quad \text{strictly proper}$$

of the polynomial equations:

$$D_c^* D_f^* z^{-g_1} G + FA_p A_q A'_{dexp} A_n = B_p^* A_r^* B_q^* B_q R_1 \tag{2.50}$$

$$D_c^* D_f^* z^{-g_1} H - FB_p A_r A_q A'_{dexp} A_n = A_p^* R_2 \tag{2.51}$$

where:

$$R_1 = z^{-g_1} (D_f D_f^* - C_n \sigma_n C_n^* A'_{dexp} A'_{dexp} A_p A_p^*) \tag{2.52}$$

$$R_2 = z^{-g_1} (D_f D_f^* A_q A_q^* B_r B_r^* + B_p B_p^* A_r A_r^* B_q B_q^* C_n \sigma_n C_n^* A'_{dexp} A'_{dexp}) \tag{2.53}$$

and $g_1 > 0$ is the smallest integer which makes the equations

(2.50)-(2.51) polynomial in d.

(ii) <u>Optimal feedforward controller</u>

$$C_{ff} = \frac{XA_r D_f - C_{cn} C_x D_{fd}}{D_{fd} A_x C_{cd}} \tag{2.54}$$

where X (along with Z and Y) is the solution having the

property:

$$(D_c^* z^{-g_2})^{-1} Z \quad \text{strictly proper}$$

of the polynomial equations:

$$D_c^* z^{-g2} X + Z A_q A_\ell A_x = z^{-g2} B_p^* A_r^* B_q^* B_q C_x D_{fd} \qquad (2.55)$$

$$D_c^* z^{-g2} Y - Z B_p A_r A_{\ell x}' = z^{-g2} A_p^* A_q^* B_r^* B_r A_{p\ell x}' C_x D_{fd} \qquad (2.56)$$

and $g_2 > 0$ is the smallest integer which makes the equations
(2.55)-(2.56) polynomial in d.

The associated minimal cost is given by:

$$J_{min} = \frac{1}{2\pi j} \oint_{|z|=1} \left[\sum_{i=1}^{2} (T_i^- T_i^{-*}) + \phi_{ol} \right] \frac{dz}{z} \qquad (2.57)$$

where the terms T_i^-, $i = \{1,2\}$ and ϕ_{ol} are defined in Appendix 1.

Proof:

The proof of Theorem 1 is given in Appendix 1.

Corollary 1

The polynomials G and H in equations (2.50) and (2.51) also
satisfy the implied cascade diophantine equation:

$$A_p H + B_p A_r G = D_f D_c \qquad (2.58)$$

which also defines the closed-loop characteristic equation i.e. :

$$\alpha = D_f D_c \qquad (2.59)$$

which, by the definitions of D_f and D_c, is a Hurwitz polynomial.

Proof:

From equation (2.34) the characteristic equation is given by:

$$\alpha = A_p C_{cd} + B_p C_{cn}$$

From equation (2.49) obtain:

$$\alpha = A_p H + B_p A_r G$$

Multiplying equation (2.50) by $B_p A_r$, equation (2.51) by A_p and then adding obtain, using equation (2.46) and cancelling common factors:

$$A_p H + B_p A_r G = D_f D_c$$

Corollary 2

The polynomials X and Y in equations (2.55) and (2.56) also satisfy the <u>implied feedforward diophantine equation</u>:

$$D_{p\ell x} A_q Y + B_p A_r X = D_c C_x D_{fd} \qquad (2.60)$$

Proof:

Multiplying equation (2.55) by $B_p A_r$, equation (2.56) by $A_q D_{p\ell x}$ and adding results, after some algebraic manipulation, in equation (2.60).

Corollary 3

The output disturbance denominator polynomial A_n is a factor of the cascade controller numerator C_{cn}.

Proof:

The diophantine equation (2.50) may be rewritten by substituting from equation (2.47) as:

$$D_c^* D_f^* z^{-gl} G + FA_p A_q A_{dexp}' A_n = B_p^* A_r^* B_q^* B_q z^{-gl} A_{pf}' A_{pf}^* A_n A_n^* ($$
$$A_{ex}' C_d^\sigma C_d^* A_{ex}'^* + A_{xd}' E_r^\sigma E_r^* A_{xd}'^* + A_{dex}^\sigma r_n A_{dex}^* + A_{ed}' C_x^\sigma \ell_n C_x^* A_{ed}'^*)$$

Since A_n divides both the right-hand side of this equation and the second term on the left side, it must also divide the term $D_c^* D_f^* z^{-gl} G$. The term $D_c^* D_f^*$ is unstable so that A_n must divide G, and hence C_{cn}, when A_n is stable. Condition (d) in Theorem 1 ensures that any unstable factors of A_n do not divide A_p and do not therefore divide D_f^*. As a consequence, A_n must again divide G, and hence C_{cn}.

•

Corollary 4

The transfer-function C_{ffn}/A_x is asymptotically stable. •

Proof:

From equation (2.54) this transfer-function may be written:

$$\frac{C_{ffn}}{A_x} = \frac{XA_r D_f - C_{cn}C_x D_{fd}}{A'_x D_{px}}$$

where D_{px} denotes any common factors of A_x and A_p such that:

$$A_x = A'_x D_{px}, \quad A_p = A'_{px} D_{px}$$

Substituting from equations (2.58) and (2.60) the above transfer-function may be written, after some algebraic manipulation, as:

$$\frac{C_{ffn}}{A_x} = \frac{C_x D_{fd} A'_{px} C_{cd} - A_q D'_{p\ell x} Y D_f}{A'_x B_p}$$

where $D'_{p\ell x} = D_{p\ell x}/D_{px}$.

Multiplying equation (2.50) by $C_x D_{fd} z^{-g2}$, equation (2.55) by R_1, and comparing obtain, after some algebraic manipulation:

$$\frac{C_{ffn}}{A_x} = (D_c^* X C_n \sigma_n C_n^* A'_{dexp} A_{dexp}^* A'_{px} A_p^* + F A'_{px} A_q A'_{dexp} A_n C_x D_{fd} z^{g1}$$

$$- Z A_q A_\ell A'_x R_1 z^{(g1+g2)}) A_r / A'_x D_c^* D_f^*$$

Comparing the above three expressions for C_{ffn}/A_x the following conclusion can be drawn : since $D_f^* D_c^*$ is strictly unstable and since

D_{px} and B_p do not have any unstable common factors (A_p and B_p cannot have unstable common factors) the expression $(XA_r D_f - C_{cn} C_x D_{fd})/D_{px}$ is, in fact, __polynomial__. By virtue of condition (c) in Theorem 1 A'_x is stable so that C_{ffn}/A_x is, as a result, asymptotically stable.

Corollary 5

Any __Hurwitz__ zeros of A_p which are not also zeros of A_d, A_x or A_e will be zeros of D_f and G (and consequently of C_{cn} and C_{ffn}). Such zeros of A_p are therefore cancelled by the controller.

Proof:

Any zero of A_p which is not also a zero of A_d, A_x or A_e will, from equation (2.19), be a zero of A'_{pf}. From equation (2.47), any Hurwitz zero of A'_{pf} will also be a zero of D_f. Denoting such a zero by A'_{pfh} then $(A'_{pfh})^2$ will be a factor of the right-hand-side of equation (2.50). Since A'_{pfh} appears in D_f^* and A_p equation (2.50) will be satisfied when A'_{pfh} appears in G (and F). From equations (2.49) and (2.54) A'_{pfh} will also be a factor of C_{cn} and C_{ffn}.

Remark

Corollary 5 includes the important limiting case when A_p has a zero on the unit circle __and__ A_p, A_x and A_e do not. Such a zero is cancelled and, although the optimal control problem may still have a solution, the closed-loop system is __not__ internally stable (see the following section). In this case the internal instability is manifest since the unstable zero appears as a factor of D_f (D_f is a factor of the closed-loop characteristic equation).

2.4.1 Internal stability

Theorem 2

The closed-loop system for the SDF plus feedforward control law is internally stable if and only if the polynomial A_p does not have any zeros on the unit circle which are not also zeros of A_d, A_x or A_e.

Proof:

The necessary and sufficient conditions for internal stability of a feedback control system derived by Kučera (1979) require that the controller can be written in the form:

$$C = X_c^{-1} Z_c$$

where X_c and Z_c are asymptotically stable transfer-functions which satisfy the Bezout identity:

$$A_p X_c + B_p Z_c = 1$$

From equation (2.49) the cascade controller may be written in the form:

$$C_c = \left(\frac{H}{D_f D_c}\right)^{-1} \left(\frac{GA_r}{D_f D_c}\right)$$

Define:

$$X_c = \frac{H}{D_f D_c} \quad , \quad Z_c = \frac{GA_r}{D_f D_c}$$

From equation (2.58) X_c and Z_c satisfy the above Bezout identity. The above condition for internal stability therefore reduces to the requirement that D_f and D_c are strictly Hurwitz polynomials. From

Lemma 1 D_c is strictly Hurwitz. From Lemma 2 D_f is strictly Hurwitz only when A_p does not have any zeros on the unit circle which are not also zeros of A_d, A_x or A_e.

In addition to the above requirements on that part of the controller present in the feedback loop, the transfer-function of the feedforward controller which is external to the loop must be asymptotically stable. From equation (2.54) the feedforward controller is given by:

$$C_{ff} = \frac{C_{ffn}}{D_{fd}A_x C_{cd}}$$

The term $1/C_{cd}$ is not necessarily stable and must in practice be included in the feedback loop (this point is discussed in detail in Section 4.6.1). The requirement therefore reduces to the stability of $C_{ffn}/D_{fd}A_x$. From Lemma 1 D_{fd} is strictly Hurwitz and from Corollary 4 the transfer-function C_{ffn}/A_x is asymptotically stable.

2.4.2 Equation solvability

To establish the solvability conditions for the cascade and feedforward diophantine equations consider firstly the general couple of equations:

$$\bar{D}X + ZM = L \tag{2.61}$$

$$\bar{D}Y - ZN = K \tag{2.62}$$

where the unknowns are X,Y and Z. Rewrite the above couple of equations in the matrix form:

$$[E] \begin{bmatrix} X \\ Y \\ Z \end{bmatrix} = [F] \tag{2.63}$$

where:

$$E \triangleq \begin{bmatrix} \bar{D} & 0 & M \\ 0 & \bar{D} & -N \end{bmatrix}, \quad F \triangleq \begin{bmatrix} L \\ K \end{bmatrix} \tag{2.64}$$

A standard result for the system of equations (2.63) is Frobenius' Theorem which states that these equations have a solution iff:

$$\text{rank } E = \text{rank}[E,F] \tag{2.65}$$

<=> the matrices E and [E,F] have the same greatest common divisors of all 1x1 and 2x2 minors.

The 1x1 and 2x2 minors of E and [E,F] are, from the definition (2.64):

1x1 minors of E : $\bar{D}$, M,N

1x1 minors of [E,F] : $\bar{D}$,M,N,L,K

2x2 minors of E : $\bar{D}^2$,$\bar{D}$N, $\bar{D}$M

2x2 minors of [E,F] : $\bar{D}^2$,$\bar{D}$N,$\bar{D}$M,$\bar{D}$K,$\bar{D}$L,MK + NL

Thus, the conditions for solvability reduce to:

(i) $(\bar{D},M,N)/L,K$

(ii) $(\bar{D}^2,\bar{D}N,\bar{D}M)/MK+NL$

where (,) denotes the greatest common divisor and / denotes divisor.

Since $(\bar{D}^2,\bar{D}N,\bar{D}M) = \bar{D}(\bar{D},M,N)$ the above two conditions reduce to:

(a) $(\bar{D},M,N)/L,K$ (2.66)

(b) $\bar{D}/MK+NL$ (2.67)

The above material relating to the couple of equations (2.61)-(2.62) is taken from Šebek (1987). Solvability of the particular couples (2.50)-(2.51) and (2.55)-(2.56) may now be investigated using conditions (a) and (b) above:

Theorem 3

When the optimal control problem solvability conditions are satisfied then the cascade controller diophantine equations (2.50)-(2.51) are solvable.

Proof:

Comparing the couples of equations (2.50)-(2.51) and (2.61)-(2.62) the polynomials $\bar{D}, M, N, L$ and K may be identified as:

$$\bar{D} = D_c^* D_f^* z^{-gl}$$

$$M = A_p A_q A'_{dexp} A_n$$

$$N = B_p A_r A_q A'_{dexp} A_n$$

$$L = B_p A_r^* B_q^* B_q R_1^*$$

$$K = A_p^* R_2$$

By definition, A_p and B_p can have no unstable common factors and A_q and A_r are strictly Hurwitz. In addition, when the problem is solvable the conditions (a)-(c) in Theorem 1 ensure that A'_{dexp} is strictly Hurwitz. Thus, the greatest common divisor of $\bar{D}$, M and N is:

$$(\bar{D}, M, N) = (D_c^* D_f^* z^{-gl}, A_n)$$

Since D_c is strictly Hurwitz and A_n has, by definition, no zeros inside the unit circle in the d-plane the only possible common factor of the polynomials $D_c^* D_f^* z^{-gl}$ and A_n is when D_f and A_n both have a zero on the unit circle. From Lemma 2 such a zero will appear in D_f^* only when A_p has a zero on the unit circle which is not also a zero of A_d, A_x or A_e. In this case, however, condition (d) in Theorem 1 ensures that such a zero <u>cannot</u> also be in A_n. As a result:

$$(\bar{D}, M, N) = 1$$

and the solvability condition (a) (2.66) is satisfied.

Finally, from the definitions above, and using equation (2.46), obtain after some algebraic manipulation:

$$MK + NL = (D_c^* D_f^* z^{-g1}) D_f D_c A_q A_{dexp}' A_n$$

and the condition (b) (2.67) is satisfied.

Theorem 4

When the optimal control problem solvability conditions are satisfied then the feedforward controller diophantine equations (2.55)-(2.56) are solvable.

Proof:

Comparing the couples of equations (2.55)-(2.56) and (2.61)-(2.62) the polynomials $\bar{D}, M, N, L$ and K may be identified as:

$$\bar{D} = D_c^* z^{-g2}$$

$$M = A_q A_\ell A_x$$

$$N = B_p A_r A_{\ell x}'$$

$$L = z^{-g2} B_p^* A_r^* B_q^* B_q C_x D_{fd}$$

$$K = z^{-g2} A_p A_q^* B_r^* B_r A_{p\ell x}' C_x D_{fd}$$

From condition (c) in Theorem 1 any unstable factors of $A_\ell A_x$ must also be in A_p and cannot, therefore, be in B_p. Condition (c) also ensures that $A_{\ell x}'$ is strictly Hurwitz, as are A_q and A_r. From Lemma 1 D_c is strictly Hurwitz and D_c^* is therefore strictly non-Hurwitz. As a consequence:

$$(\bar{D}, M, N) = 1$$

and condition (a) (2.66) is satisfied.

From the above definitions, and using equations (2.27) and (2.46) obtain after some algebraic manipulation:

$$MK + NL = (z^{-g2}D_c^*)D_c C_x D_{fd}A'_{\ell x}$$

and the condition (b) (2.67) is seen to be satisfied.

2.4.3 Zero output-measurement noise

When the measurement noise $n(t)$ acting on the controlled output $y(t)$ is zero the diophantine equations for the cascade controller simplify as follows:

Theorem 5

When the measurement noise $n(t) = 0$ the spectral factor D_f is defined by:

$$D_f D_f^* = (A'_{ex}C_d \sigma_d C_d^* A'^{*}_{ex} + A'_{xd}E_r \sigma_r E_r^* A'^{*}_{xd}$$
$$+ A_{dex}\sigma_{rn}A_{dex}^* + A'_{ed}C_x \sigma_{\ell n}C_x^* A'^{*}_{ed})A'_{pf}A'^{*}_{pf} \qquad (2.68)$$

The optimal cascade controller is given by:

$$C_c = \frac{GA_r}{H'A_q} \qquad (2.69)$$

where G, H' (along with F') is the solution having the property:

$$(D_c^* z^{-g1})^{-1}F' \quad \text{strictly proper}$$

of the polynomial equations:

$$D_c^* z^{-g1}G + F'A_p A_q A'_{dexp} = z^{-g1}B_p^* A_r^* B_q^* B_q D_f \qquad (2.70)$$

$$D_c^* z^{-gl} H' - F'B_p A_r A'_{dexp} = z^{-gl} A_p^* A_q^* B_r^* B_r D_f \tag{2.71}$$

where $gl > 0$ is the smallest integer which makes the equations (2.70)-(2.71) polynomial in d.

The polynomials G and H' in equations (2.70)-(2.71) also satisfy the diophantine equation:

$$A_p A_q H' + B_p A_r G = D_f D_c \tag{2.72}$$

Proof:

To obtain $n(t) = 0$ set $\sigma_n = C_n = 0$, $A_n = 1$. The definition of D_f in equation (2.68) then follows immediately from equation (2.47).

The diophantine equation (2.50) becomes:

$$D_c^* D_f^* z^{-gl} G + FA_p A_q A'_{dexp} = z^{-gl} B_p^* A_r^* B_q^* B_q D_f D_f^*$$

Since D_f^* divides the right-hand-side and the first term on the left side of this equation, it must also divide F. Denoting F by:

$$F = D_f^* F'$$

and cancelling the common factor D_f^* results in equation (2.70).

The diophantine equation (2.51) becomes:

$$D_c^* D_f^* z^{-gl} H - FB_p A_r A_q A'_{dexp} = z^{-gl} A_p^* A_q^* B_r^* B_r D_f A_q D_f^*$$

By a similar reasoning to that used above, D_f^* must again divide F. Since A_q divides the right-hand-side and the second term on the left side of this equation, it must also divide H. Denoting H by:

$$H = A_q H'$$

and cancelling the common factor $A_q D_f^*$ results in equation (2.71).

Using equation (2.49) the cascade controller becomes:

$$C_c = \frac{GA_r}{H'A_q}$$

Since $F = D_f^* F'$ the conditions $(D_c^* D_f^* z^{-g1})^{-1} F$ strictly proper and $(D_c^* z^{-g1})^{-1} F'$ strictly proper are clearly equivalent.

Multiplying equation (2.70) by $B_p A_r$, equation (2.71) by $A_p A_q$ and then adding results, using equation (2.46) and cancelling common factors, in equation (2.72).

2.5 TWO-DEGREES-OF-FREEDOM SOLUTION WITH FEEDFORWARD

The Hurwitz spectral factors D_c, D_f, D_{fd} and D_m are defined by:

$$D_c D_c^* = B_p A_r B_q B_q^* A_r^* B_p^* + A_p A_q B_r B_r^* A_q^* A_p^* \tag{2.73}$$

$$D_f D_f^* = (A_n A_x' C_d \sigma_d C_d^* A_x'^* A_n^* + A_{dx} C_n \sigma_n C_n^* A_{dx}^* + A_d' A_n C_x \sigma_{\ell n} C_x^* A_n^* A_d'^*) A_p' A_p'^* \tag{2.74}$$

$$D_{fd} D_{fd}^* = A_\ell \sigma_{\ell n} A_\ell^* + E_\ell \sigma_\ell E_\ell^* \tag{2.75}$$

$$D_m D_m^* = A_e \sigma_{rn} A_e^* + E_r \sigma_r E_r^* \tag{2.76}$$

Lemma 1

The polynomials D_c, D_{fd} and D_m are strictly Hurwitz.

Proof:

For D_c and D_{fd} the proof is the same as that of Lemma 1 in Section 2.4.

Any unstable zero of D_m would, from equation (2.76), require:

$$\sigma_{rn} |A_e(e^{j\omega})|^2 + \sigma_r |E_r(e^{j\omega})|^2 = 0$$

Since the intensities σ_{rn} and σ_r are non-zero this condition would require that $d = \exp(j\omega)$ is a zero of both A_e and E_r which is not possible (Assumption (i) in Section 2.1). As a consequence D_m is strictly Hurwitz.

Lemma 2

Any zeros of A_p which lie on the unit circle and which are not also zeros of A_d or A_x will be zeros of the spectral factor D_f. If no such zeros exist then D_f is strictly Hurwitz.

Proof:

Any zero of A_p which is not also a zero of A_d or A_x will, from equation (2.12), be a zero of A_p'. Any such zero lying on the unit circle will, from equation (2.74), also be a zero of D_f.

Since C_d, C_n and C_x have no common factors on the unit circle (Assumption (iv) in Section 2.1) the term inside the brackets in equation (2.74) can only lead to strictly Hurwitz terms in D_f.

Theorem 6 : Optimal 2DF plus feedforward controller

The optimal control problem for the 2DF control structure has a solution if and only if:

(a)   Any unstable factors of A_d are also factors of A_p.

(b)   Any unstable factors of A_e are also factors of A_p.

(c)   Any unstable factors of the product $A_\ell A_x$ are also factors of A_p.

(d)   Any unstable factors of A_n are <u>not</u> also factors of A_p.

The feedback, reference and feedforward parts of the control law (2.35) which minimises the cost-function (2.43) are as follows:

(i)   Optimal feedback controller

$$C_{fb} = \frac{GA_r}{H} \qquad\qquad (2.77)$$

where G,H (along with F) is the solution having the property:

$$(D_c^* D_f^* z^{-g1})^{-1} F \quad \text{strictly proper}$$

of the polynomial equations:

$$D_c^* D_f^* z^{-g_1} G + FA_p A_q A'_{dx} A_n = B_p^* A_r^* B_q^* B_q R_1 \qquad (2.78)$$

$$D_c^* D_f^* z^{-g_1} H - FB_p A_r A_q A'_{dx} A_n = A_p^* R_2 \qquad (2.79)$$

where:

$$R_1 = z^{-g_1}(D_f D_f^* - C_n \sigma_n C_n^* A'_{dx} A_{dx}^* A_p A_p^*) \qquad (2.80)$$

$$R_2 = z^{-g_1}(D_f D_f^* A_q A_q^* B_r B_r^* + B_p B_p^* A_r A_r^* A_q B_q B_q^* C_n \sigma_n C_n^* A'_{dx} A_{dx}^*) \qquad (2.81)$$

and $g_1 > 0$ is the smallest integer which makes the equations (2.78)-(2.79) polynomial in d.

(ii) <u>Optimal reference controller</u>

$$C_r = \frac{MA_r D_f}{D_m C_{fbd}} \qquad (2.82)$$

where M (along with N and Q) is the solution having the property:

$$(D_c^* z^{-g_2})^{-1} N \quad \text{strictly proper}$$

of the polynomial equations:

$$D_c^* z^{-g_2} M + NA_q A_e = z^{-g_2} B_p^* A_r^* B_q^* B_q D_m \qquad (2.83)$$

$$D_c^* z^{-g_2} Q - NB_p A_r A'_e = z^{-g_2} A_p^* A_q^* B_r^* B_q B_r A'_{pe} D_m \qquad (2.84)$$

and g2 > 0 is the smallest integer which makes the equations (2.83)-(2.84) polynomial in d.

(iii) <u>Optimal feedforward controller</u>

$$C_{ff} = \frac{XA_r D_f - C_{fbn} C_x D_{fd}}{D_{fd} A_x C_{fbd}} \qquad (2.85)$$

where X (along with Z and Y) is the solution having the property:

$$(D_c^* z^{-g_3})^{-1} Z \quad \text{strictly proper}$$

of the polynomial equations:

$$D_c^* z^{-g3} X + ZA_q A_\ell A_x = z^{-g3} B_p^* A_r^* B_q^* C_x D_{fd}$$ (2.86)

$$D_c^* z^{-g3} Y - ZB_p A_r A'_{\ell x} = z^{-g3} A_p^* A_q^* B_r^* B_r A'_{p\ell x} C_x D_{fd}$$ (2.87)

and g3 > 0 is the smallest integer which makes the equations (2.86)-(2.87) polynomial in d.

The associated minimal cost is given by:

$$J_{min} = \frac{1}{2\pi j} \oint_{|z|=1} \left[\sum_{i=1}^{3} (T_i^- T_i^{-*}) + \phi_{ol} \right] \frac{dz}{z}$$ (2.88)

where the terms T_i^-, $i = \{1,2,3\}$ and ϕ_{ol} are defined in Appendix 2.

Proof:

The proof of Theorem 6 is given in Appendix 2.

Corollary 1

The polynomials G and H in equations (2.78) and (2.79) also satisfy the implied feedback diophantine equation:

$$A_p H + B_p A_r G = D_f D_c$$ (2.89)

which also defines the closed-loop characteristic equation i.e.:

$$\alpha = D_f D_c$$ (2.90)

which, by the definitions of D_f and D_c, is a Hurwitz polynomial.

Proof

From equation (2.41) the characteristic equation is given by:

$$\alpha = A_p C_{fbd} + B_p C_{fbn}$$

From equation (2.77) obtain:

$$\alpha = A_p H + B_p A_r G$$

Multiplying equation (2.78) by $B_p A_r$, equation (2.79) by A_p and then adding obtain, using equation (2.73) and cancelling common factors:

$$A_p H + B_p A_r G = D_f D_c$$

Corollary 2

The polynomials M and Q in equations (2.83) and (2.84) also satisfy the implied reference diophantine equation:

$$D_{pe} A_q Q + B_p A_r M = D_c D_m \qquad (2.91)$$

Proof:

Multiplying equation (2.83) by $B_p A_r$, equation (2.84) by $A_q D_{pe}$ and adding results, after some algebraic manipulation, in equation (2.91).

Corollary 3

The polynomials X and Y in equations (2.86) and (2.87) also satisfy the implied feedforward diophantine equation:

$$D_{p\ell x} A_q Y + B_p A_r X = D_c C_x D_{fd} \qquad (2.92)$$

Proof:

Multiplying equation (2.86) by $B_p A_r$, equation (2.87) by $A_q D_{p\ell x}$ and adding results, after some algebraic manipulation, in equation (2.92).

Corollary 4

The output disturbance denominator polynomial A_n is a factor of the feedback controller numerator C_{fbn}.

Proof:

The diophantine equation (2.78) may be rewritten by substituting from equation (2.74) as:

$$D_c^* D_f^* z^{-gl} G + FA_p A_q A_{dx}' A_n = B_p^* A_r^* B_q B_q^* z^{-gl} A_p' A_p'^* A_n A_n^* ($$
$$A_x' C_d \sigma_d C_d^* A_x'^* + A_d' C_x \sigma_{\ell n} C_x^* A_d'^*)$$

Since A_n divides both the right-hand-side of this equation and the second term on the left side, it must also divide the term $D_c^* D_f^* z^{-gl} G$. The term $D_c^* D_f^*$ is unstable so that A_n must divide G, and hence C_{fbn}, when A_n is stable. Condition (d) in Theorem 6 ensures that any unstable factors of A_n do not divide A_p and do not therefore divide D_f^*. As a consequence, A_n must again divide G, and hence C_{fbn}.

Corollary 5

The transfer-function C_{ffn}/A_x is asymptotically stable.

Proof:

From equation (2.85) this transfer-function may be written:

$$\frac{C_{ffn}}{A_x} = \frac{X A_r D_f - C_{fbn} C_x D_{fd}}{A_x' D_{px}}$$

where D_{px} denotes any common factors of A_x and A_p such that:

$$A_x = A_x' D_{px}, \quad A_p = A_{px}' D_{px}$$

Substituting from equations (2.89) and (2.92) the above transfer-function may be written, after some algebraic manipulation, as:

$$\frac{C_{ffn}}{A_x} = \frac{C_x D_{fd} A'_{px} C_{fbd} - A_q D'_{p\ell x} Y D_f}{A'_x B_p}$$

where $D'_{p\ell x} = D_{p\ell x}/D_{px}$.

Multiplying equation (2.78) by $C_x D_{fd} z^{-g3}$, equation (2.86) by R_1 and comparing obtain, after some algebraic manipulation:

$$\frac{C_{ffn}}{A_x} = (D_c^* X C_n \sigma_n C^* A'_n A_{dx}^{*} A'_{dx} A'_{px} A_p^* + F A'_{px} A_q A'_{dx} A_n C_x D_{fd} z^{g1}$$

$$- Z A_q A_\ell A'_x R_1 z^{(g1+g3)}) A_r / A'_x D_f^* D_c^*$$

Comparing the above three expressions for C_{ffn}/A_x the following conclusion may be drawn : since $D_f^* D_c^*$ is strictly unstable and since D_{px} and B_p do not have any unstable common factors (A_p and B_p cannot have any unstable common factors) the expression

$(X A_r D_f - C_{fbn} C_x D_{fd})/D_{px}$ is, in fact, <u>polynomial</u>. By virtue of condition (c) in Theorem 6 A'_x is stable so that C_{ffn}/A_x is, as a result, asymptotically stable.

Corollary 6

Any Hurwitz zeros of A_p which are not also zeros of A_d or A_x will be zeros of D_f and G (and consequently of C_{fbn}, C_{rn} and C_{ffn}). Such zeros of A_p are therefore cancelled by the controller.

Proof:

Any zero of A_p which is not a zero of A_d or A_x will, from equation (2.12), be a zero of A'_p. From equation (2.74), any Hurwitz zero of A'_p will also be a zero of D_f. Denoting such a zero by A'_{ph} then $(A'_{ph})^2$ will be a factor of the right-hand-side of equation (2.78). Since A'_{ph} appears in D_f^* and A_p, equation (2.78) will be satisfied when A'_{ph} appears in G (and F). From equations (2.77), (2.82) and (2.85) A'_{ph} will also be a factor of C_{fbn}, C_{rn} and C_{ffn}.

Remark

Corollary 6 includes the important limiting case when A_p has a zero on the unit circle <u>and</u> A_p and A_x do not. Such a zero is cancelled and, although the optimal control problem may still be solvable, the closed-loop system is not internally stable (see the following section).

2.5.1 Internal stability

Theorem 7

The closed-loop system for the 2DF plus feedforward control law is internally stable if and only if the polynomial A_p does not have any zeros on the unit circle which are not also zeros of A_d or A_x.

Proof:

From equation (2.77) the feedback controller may be written in the form:

$$C_{fb} = (\frac{H}{D_f D_c})^{-1} (\frac{GA_r}{D_f D_c})$$

Define:

$$X_c = \frac{H}{D_f D_c} \quad , \quad Z_c = \frac{GA_r}{D_f D_c}$$

From equation (2.89) X_c and Z_c satisfy the above Bezout identity. Following the conditions stated in the proof of Theorem 2 the polynomials D_c and D_f are required to be strictly Hurwitz for the

system to be internally stable. From Lemma 1 D_c is strictly Hurwitz.
From Lemma 2 D_f is strictly Hurwitz only when A_p does not have any
zeros on the unit circle which are not also zeros of A_d or A_x.

In addition to the above requirements on that part of the
controller present in the feedback loop, the transfer-functions of
the reference and feedforward controllers which are external to the
loop must be asymptotically stable.

From equation (2.82) the reference controller is given by:

$$C_r = \frac{MA_r D_f}{D_m C_{fbd}}$$

The term $1/C_{fbd}$ is not necessarily stable and must in practice be
included in the feedback loop (this point is discussed in detail in
Section 4.6.1). The requirement therefore reduces to the stability
of $MA_r D_f/D_m$. From Lemma 1 D_m is strictly Hurwitz.

From equation (2.85) the feedforward controller is given by:

$$C_{ff} = \frac{C_{ffn}}{D_{fd} A_x C_{fbd}}$$

The term $1/C_{fbd}$ must again be included in the feedback loop so that
only $C_{ffn}/D_{fd} A_x$ must be asymptotically stable. By Lemma 1 D_{fd} is
strictly Hurwitz. By Corollary 5 C_{ffn}/A_x is asymptotically stable.

2.5.2 Equation solvability

The solvability conditions for the feedback, reference and
feedforward diophantine equations are established using the general
theory outlined in Section 2.4.2:

57

<u>Theorem 8</u>

When the optimal control problem solvability conditions are satisfied then the feedback controller diophantine equations (2.78)-(2.79) are solvable.

<u>Proof</u>

Comparing the couples of equations (2.61)-(2.62) and (2.78)-(2.79) the polynomials $\bar{D}, M, N, L$ and K may be identified as:

$$\bar{D} = z^{-g1} D_c^* D_f^*$$

$$M = A_p A_q A_{dx}' A_n$$

$$N = B_p A_r A_q A_{dx}' A_n$$

$$L = B_p A_r^* B_q^* B_q R_1$$

$$K = A_p^* R_2$$

By definition A_p and B_p can have no unstable common factors and A_q and A_r are strictly Hurwitz. In addition, when the problem is solvable the conditions (a) and (c) in Theorem 6 ensure that A_{dx}' is strictly Hurwitz. Thus, the greatest common divisor of $\bar{D}, M$ and N is:

$$(\bar{D}, M, N) = (D_c^* D_f^* z^{-g1}, A_n)$$

Since D_c is strictly Hurwitz and A_n has, by definition, no zeros inside the unit circle in the d-plane the only possible common factor of the polynomials $D_c^* D_f^* z^{-g1}$ and A_n is when D_f and A_n both have a zero on the unit circle. From Lemma 2 such a zero will appear in D_f only when A_p has a zero on the unit circle which is not also a zero of A_d or A_x. In this case, however, condition (d) in Theorem 6 ensures that such a zero <u>cannot</u> also be in A_n. As a result:

$$(\bar{D}, M, N) = 1$$

and the solvability condition (a) (2.66) is satisfied. Finally, from the above definitions, and using equation (2.73), obtain after some algebraic manipulation:

$$MK + NL = (D_c^* D_f^* z^{-g1}) D_f D_c A_q A_{dx}' A_n$$

and the condition (b) (2.67) is satisfied.

Theorem 9

When the optimal control problem solvability conditions are satisfied the reference controller diophantine equations (2.83)-(2.84) are solvable.

Proof

Comparing the couples of equations (2.61)-(2.62) and (2.83)-(2.84) the polynomials $\bar{D}, M, N, L$ and K may be identified as:

$$\bar{D} = D_c^* z^{-g2}$$
$$M = A_q A_e$$
$$N = B_p A_r A_e'$$
$$L = z^{-g2} B_p^* A_r B_q^* B_q^* D_m$$
$$K = z^{-g2} A_p^* A_q B_r^* B_r A_{pe}' D_m$$

From condition (b) in Theorem 6 any unstable factors of A_e must also be in A_p and cannot, therefore, be in B_p. Condition (b) also ensures that A_e' is strictly Hurwitz, as are A_q and A_r. From Lemma 1 D_c is strictly Hurwitz and D_c^* is therefore strictly non-Hurwitz. As a consequence:

$$(\bar{D}, M, N) = 1$$

and condition (a) (2.66) is satisfied.

From the above definitions, and using equations (2.17) and (2.73), obtain after some algebraic manipulation:

$$MK + NL = (z^{-g2}D_c^*)D_c A_e' D_m$$

and the condition (b) (2.67) is seen to be satisfied.

Theorem 10

When the optimal control problem solvability conditions are satisfied then the feedforward controller diophantine equations (2.86)-(2.87) are solvable.

Proof

The couples of equations (2.55)-(2.56) and (2.86)-(2.87) are identical. The proof then follows by analogy with the proof of Theorem 4.

2.5.3 Zero output-measurement noise

When the measurement noise n(t) acting on the controlled output y(t) is zero the diophantine equations for the feedback controller simplify as follows:

Theorem 11

When the measurement noise n(t) = 0 the spectral factor D_f is defined by:

$$D_f D_f^* = (A_x' C_d \sigma_d C_d^* A_x'^* + A_d' C_x \sigma_{\ell n} C_x^* A_d'^*)A_p' A_p'^* \tag{2.93}$$

The optimal feedback controller is given by:

$$C_{fb} = \frac{GA_r}{H'A_q} \tag{2.94}$$

where G,H' (along with F') is the solution having the property:

$(D_c^* z^{-gl})^{-1} F'$ strictly proper

of the polynomial equations:

$$D_c^* z^{-gl} G + F'A_p A_q A'_{dx} = z^{-gl} B_p^* A_r^* B_q^* B_q D_f \tag{2.95}$$

$$D_c^* z^{-gl} H' - F'B_p A_r A'_{dx} = z^{-gl} A_p^* A_q^* A_r^* B_r B_q D_f \tag{2.96}$$

where gl > 0 is the smallest integer which makes the equations (2.95)-(2.96) polynomial in d.

The polynomials G and H' in equations (2.95)-(2.96) also satisfy the diophantine equation:

$$A_p A_q H' + B_p A_r G = D_f D_c \tag{2.97}.$$

Proof:

To obtain n(t) = 0 set $\sigma_n = C_n = 0$, $A_n = 1$. The definition of D_f in equation (2.93) then follows immediately from equation (2.74).

The diophantine equation (2.78) becomes:

$$D_c^* D_f^* z^{-gl} G + FA_p A_q A'_{dx} = z^{-gl} B_p^* A_r^* B_q^* B_q D_f D_f^*$$

Since D_f^* divides the right-hand-side and the first term on the left side of this equation, it must also divide F. Denoting F as:

$$F = D_f^* F'$$

and cancelling the common factor D_f^* results in equation (2.95).

The diophantine equation (2.79) becomes:

$$D_c^* D_f^* z^{-gl} H - FB_p A_r A_q A'_{dx} = z^{-gl} A_p^* A_q^* A_r^* B_r B_q D_f A_q D_f^*$$

By a similar reasoning to that used above, D_f^* must again divide F.

Since A_q divides the right-hand-side and the second term on the left side of this equation, it must also divide H. Denoting H as:

$$H = A_q H'$$

and cancelling the common factor $A_q D_f^*$ results in equation (2.96).

Using equation (2.77) the feedback controller equation becomes:

$$C_{fb} = \frac{GA_r}{H'A_q}$$

Since $F = D_f^* F'$ the conditions $(D_c^* D_f^* z^{-gl})^{-1} F$ strictly proper, and $(D_c^* z^{-gl})^{-1} F'$ strictly proper, are clearly equivalent.

Multiplying equation (2.95) by $B_p A_r$, equation (2.96) by $A_p A_q$ and then adding results, using equation (2.73) and cancelling common factors, in equation (2.97).

2.6 SDF SOLUTION USING A COMMON DENOMINATOR

In the single-degree-of-freedom control structure shown in Figure 2.3 it is always possible to express the various sub-systems using a least-common-denominator polynomial.   Denoting the least-common-factor of A_p, A_x, A_d and A_e by A i.e:

$$A \triangleq \text{l.c.m.} \ (A_p, \ A_x, \ A_d, \ A_e) \tag{2.98}$$

then the sub-systems W_p, W_x, W_d and W_r may be expressed using their least-common-denominator A as:

$$W_p = A^{-1}B \tag{2.99}$$

$$W_d = A^{-1}C \tag{2.100}$$

$$W_x = A^{-1}D \tag{2.101}$$

$$W_r = A^{-1}E \tag{2.102}$$

As before, the sub-systems W_n and W_ℓ are denoted by:

$$W_n = A_n^{-1}C_n \tag{2.103}$$

$$W_\ell = A_\ell^{-1}E_\ell \tag{2.104}$$

Theorem 12: SDF Solution Using a Common Denominator

The optimal control problem for the SDF control structure using the common denominator model (2.99)-(2.102) has a solution if and only if:

(a)   A and B have no unstable common factors.

(b)   Any unstable factors of A_ℓ are also factors of A and D.

(c)   Any unstable factors of A_n are not also factors of A.

The Hurwitz spectral factors D_c, D_f and D_{fd} are defined by:

$$D_c D_c^* = BA_r B_q B_q^* A_r^* B^* + AA_q B_r B_r^* A_q^* A^* \tag{2.105}$$

$$D_f D_f^* = (A_n C \sigma_d C^* A_n^* + A C_n \sigma_n C_n^* A^* + A_n E \sigma_r E^* A_n^*$$
$$+ AA_n \sigma_{rn} A_n^* A^* + A_n D \sigma_{\ell n} D^* A_n^*) \tag{2.106}$$

$$D_{fd} D_{fd}^* = A_\ell \sigma_{\ell n} A_\ell^* + E_\ell \sigma_\ell E_\ell^* \tag{2.107}$$

The cascade and feedforward parts of the control law (2.29) which minimises the cost-function (2.43) are as follows:

(i) <u>Optimal cascade controller</u>

$$C_c = \frac{GA_r}{H} \tag{2.108}$$

where G,H (along with F) is the solution having the property:

$$(D_c^* D_f^* z^{-g_1})^{-1} F \quad \text{strictly proper}$$

of the polynomial equations:

$$D_c^* D_f^* z^{-g_1} G + FAA_q A_n = B^* A_r^* B_q^* B_q R_1 \tag{2.109}$$

$$D_c^* D_f^* z^{-g_1} H - FBA_r A_q A_n = A^* R_2 \tag{2.110}$$

where:

$$R_1 = z^{-g_1}(D_f D_f^* - C_n \sigma_n C_n^* AA^*) \tag{2.111}$$

$$R_2 = z^{-g_1}(D_f D_f^* A_q A_q^* B_r B_r^* + BB^* A_r A_r^* B_q B_q^* C_n \sigma_n C_n^*) \tag{2.112}$$

and $g_1 > 0$ is the smallest integer which makes the equations (2.109)-(2.110) polynomial in d.

(ii) <u>Optimal feedforward controller</u>

$$C_{ff} = \frac{XA_r D_f - C_{cn} DD_{fd}}{D_{fd} AC_{cd}} \tag{2.113}$$

where X (along with Z and Y) is the solution having the property:

$$(D_c^* z^{-g2})^{-1} z \quad \text{strictly proper}$$

of the polynomial equations:

$$D_c^* z^{-g2} X + ZAA_q A_\ell = z^{-g2} B^* A_r^* B_q^* B_q DD_{fd} \qquad (2.114)$$

$$D_c^* z^{-g2} Y - ZBA_r A_\ell = z^{-g2} A^* A_q^* B_r^* B_r DD_{fd} \qquad (2.115)$$

and $g2 > 0$ is the smallest integer which makes the equations (2.114)-(2.115) polynomial in d.

The associated minimal cost is given by:

$$J_{min} = \frac{1}{2\pi j} \oint_{|z|=1} \left[\sum_{i=1}^{2} (T_i^- T_i^{-*}) + \phi_{ol} \right] \frac{dz}{z} \qquad (2.116)$$

where the terms T_i^-, $i = \{1,2\}$ and ϕ_{ol} are defined in Appendices 3 and 1, respectively.

Proof:

The proof of Theorem 12 is given in Appendix 3.

Corollary 1

The polynomials G and H in equations (2.109) and (2.110) also satisfy the implied cascade diophantine equation:

$$AH + BA_r G = D_f D_c \qquad (2.117)$$

Proof:

Multiplying equation (2.109) by BA_r, equation (2.110) by A and then adding results, using equation (2.105) and cancelling common factors, in equation (2.117).

Corollary 2

The polynomials X and Y in equations (2.114) and (2.115) also satisfy the <u>implied feedforward diophantine equation</u>:

$$AA_q Y + BA_r X = D_c DD_{fd} \qquad (2.118)$$

<u>Proof</u>:

Multiplying equation (2.114) by BA_r, equation (2.115) by AA_q and adding results, after some algebraic manipulation, in equation (2.118).

2.6.1 Zero output-measurement noise

When the measurement noise $n(t)$ acting on the controlled output $y(t)$ is zero the diophantine equations for the cascade controller simplify as follows:

<u>Theorem 13</u>

When the measurement noise $n(t) = 0$ the spectral factor D_f is defined by:

$$D_f D_f^* = C\sigma_d C^* + E\sigma_r E^* + A\sigma_{rn} A^* + D\sigma_{\ell n} D^* \qquad (2.119)$$

The optimal cascade controller is given by:

$$C_c = \frac{GA_r}{H'A_q} \qquad (2.120)$$

where G, H' (along with F') is the solution having the property:

$$(D_c^* z^{-g1})^{-1} F' \quad \text{strictly proper}$$

of the polynomial equations:

$$D_c^* z^{-g1} G + F'AA_q = z^{-g1} B_r A_q^* B_q^* B_q D_f \qquad (2.121)$$

$$D_c^* z^{-g1} H' - F'BA_r = z^{-g1} A_q^* A B_r^* B_r D_f \qquad (2.122)$$

where $g1 > 0$ is the smallest integer which makes the equations (2.121)-(2.122) polynomial in d.

The polynomials G and H' in equations (2.121)-(2.122) also satisfy the diophantine equation:

$$AA_q H' + BA_r G = D_f D_c \tag{2.123}$$

Proof:

To obtain $n(t) = 0$ set $\sigma_n = C_n = 0$, $A_n = 1$. The definition of D_f in equation (2.119) then follows immediately from equation (2.106).

The diophantine equation (2.109) becomes:

$$D_c^* D_f^* z^{-g1} G + FAA_q = z^{-g1} B^* A_r^* B_q^* B_q D_f D_f^*$$

Since D_f^* divides the right-hand-side and the first term on the left side of this equation, it must also divide F. Denoting F by:

$$F = D_f^* F'$$

and cancelling the common factor D_f^* results in equation (2.121).

The diophantine equation (2.110) becomes:

$$D_c^* D_f^* z^{-g1} H - FBA_r A_q = z^{-g1} A^* A_q^* B_r^* B_r D_f A_q D_f^*$$

By a similar reasoning to that used above, D_f^* must again divide F. Since A_q divides the right-hand-side and the second term on the left side of this equation, it must also divide H. Denoting H by:

$$H = A_q H'$$

and cancelling the common factor $A_q D_f^*$ results in equation (2.122).

Using equation (2.108) the cascade controller equation becomes:

$$C_c = \frac{GA_r}{H'A_q}$$

Since $F = D_f^* F'$ the conditions $(D_c^* D_f^* z^{-g1})^{-1} F$ strictly proper and

$(D_c^* z^{-g1})^{-1} F'$ strictly proper are clearly equivalent.

Multiplying equation (2.121) by BA_r, equation (2.122) by AA_q and then adding results, using equation (2.105) and cancelling common factors, in equation (2.123).

2.7 2DF SOLUTION USING A COMMON DENOMINATOR

In the two-degrees-of-freedom control structure shown in Figure 2.4 it is always possible to express the various sub-systems using a least-common-denominator polynomial. Denoting the least-common-factor of A_p, A_d, and A_x by A i.e:

$$A \triangleq \text{l.c.m.} (A_p, A_d, A_x) \qquad (2.124)$$

then the sub-systems W_p, W_d, and W_x may be expressed using their least-common-denominator A as:

$$W_p = A^{-1}B \qquad (2.125)$$

$$W_d = A^{-1}C \qquad (2.126)$$

$$W_x = A^{-1}D \qquad (2.127)$$

As before, the sub-systems W_n, W_ℓ and W_r are denoted by:

$$W_n = A_n^{-1}C_n \qquad (2.128)$$

$$W_\ell = A_\ell^{-1}E_\ell \qquad (2.129)$$

$$W_r = A_e^{-1}E_r \qquad (2.130)$$

Any common factors of A_e and A are denoted by D_e such that:

$$A_e = D_e A'_{ec} \quad , \quad A = D_e A' \qquad (2.131)$$

Theorem 14: 2DF Solution Using a Common Denominator

The optimal control problem for the 2DF control structure using the common denominator model (2.125)-(2.127) has a solution if and only if:

(a) A and B have no unstable common factors.

(b) Any unstable factors of A_e are also factors of A.

(c)   Any unstable factors of A_ℓ are also factors of A and D.

(d)   Any unstable factors of A_n are <u>not</u> also factors of A.

The Hurwitz spectral factors D_c, D_f, D_{fd} and D_m are defined by:

$$D_c D_c^* = BA_r B_q B_q^* A_r^* B^* + AA_q B_r B_r^* A_q^* A^* \qquad (2.132)$$

$$D_f D_f^* = A_n C \sigma_d C^* A_n^* + AC_n \sigma_n C_n^* A^* + A_n D \sigma_{\ell n} D^* A_n^* \qquad (2.133)$$

$$D_{fd} D_{fd}^* = A_\ell \sigma_{\ell n} A_\ell^* + E_\ell \sigma_\ell E_\ell^* \qquad (2.134)$$

$$D_m D_m^* = A_e \sigma_{rn} A_e^* + E_r \sigma_r E_r^* \qquad (2.135)$$

The feedback, reference and feedforward parts of the control law (2.35) which minimises the cost-function (2.43) are as follows:

(i)   <u>Optimal feedback controller</u>

$$C_{fb} = \frac{GA_r}{H} \qquad (2.136)$$

where G,H (along with F) is the solution having the property:

$$(D_c^* D_f^* z^{-g_1})^{-1} F \quad \text{strictly proper}$$

of the polynomial equations:

$$D_c^* D_f^* z^{-g_1} G + FAA_q A_n = B^* A_r^* B_q^* B_q R_1 \qquad (2.137)$$

$$D_c^* D_f^* z^{-g_1} H - FBA_r A_q A_n = A^* R_2 \qquad (2.138)$$

where:

$$R_1 = z^{-g_1}(D_f D_f^* - C_n \sigma_n C_n^* AA^*) \qquad (2.139)$$

$$R_2 = z^{-g_1}(D_f D_f^* A_q A_q^* B_r B_r^* + BB^* A_r A_r^* B_q B_q^* C_n \sigma_n C_n^*) \qquad (2.140)$$

and $g_1 > 0$ is the smallest integer which makes the equations (2.137)-(2.138) polynomial in d.

(ii)   <u>Optimal reference controller</u>

$$C_r = \frac{MA_r D_f}{D_m C_{fbd}} \qquad (2.141)$$

where M (along with N and Q) is the solution having the property:

$$(D_c^* z^{-g2})^{-1} N \quad \text{strictly proper}$$

of the polynomial equations:

$$D_c^* z^{-g2} M + N A_q A_e = z^{-g2} B^* A_r^* B_q^* B_q D_m \tag{2.142}$$

$$D_c^* z^{-g2} Q - N B A_r A'_{ec} = z^{-g2} A^* A_q^* B_r^* B_r A' D_m \tag{2.143}$$

and g2 > 0 is the smallest integer which makes the equations (2.142)-(2.143) polynomial in d.

(iii) <u>Optimal feedforward controller</u>

$$C_{ff} = \frac{X A_r D_f - C_{fbn} DD_{fd}}{D_{fd} A C_{fbd}} \tag{2.144}$$

where X (along with Z and Y) is the solution having the property:

$$(D_c^* z^{-g3})^{-1} Z \quad \text{strictly proper}$$

of the polynomial equations:

$$D_c^* z^{-g3} X + Z A A_q A_\ell = z^{-g3} B^* A_r^* B_q^* B_q DD_{fd} \tag{2.145}$$

$$D_c^* z^{-g3} Y - Z B A_r A_\ell = z^{-g3} A^* A_q^* B_r^* B_r DD_{fd} \tag{2.146}$$

and g3 > 0 is the smallest integer which makes the equations (2.145)-(2.146) polynomial in d.

The associated minimal cost is given by:

$$J_{min} = \frac{1}{2\pi j} \oint_{|z|=1} \left[\sum_{i=1}^{3} (T_i^- T_i^{-*}) + \phi_{ol} \right] \frac{dz}{z} \tag{2.147}$$

where the terms T_i^-, i = {1,2,3} and ϕ_{ol} are defined in Appendices 4 and 2, respectively.

71

Proof:

The proof of Theorem 14 is given in Appendix 4.

Corollary 1

The polynomials G and H in equations (2.137) and (2.138) also satisfy the implied feedback diophantine equation:

$$AH + BA_r G = D_f D_c \qquad (2.148)$$

Proof:

Multiplying equation (2.137) by BA_r, equation (2.138) by A and then adding results, using equation (2.132) and cancelling common factors, in equation (2.148).

Corollary 2

The polynomials M and Q in equations (2.142) and (2.143) also satisfy the implied reference diophantine equation:

$$D_e A_q Q + BA_r M = D_c D_m \qquad (2.149)$$

Proof:

Multiplying equation (2.142) by BA_r, equation (2.143) by $D_e A_q$ and adding results, using equation (2.132) and cancelling common factors, in equation (2.149).

Corollary 3

The polynomials X and Y in equations (2.145) and (2.146) also satisfy the implied feedforward diophantine equation:

$$AA_q Y + BA_r X = D_c DD_{fd} \qquad (2.150)$$

Proof:

Multiplying equation (2.145) by BA_r, equation (2.146) by AA_q and adding results, using equation (2.132) and cancelling common factors, in equation (2.150).

2.7.1 Zero output-measurement noise

When the measurement noise $n(t)$ acting on the controlled output $y(t)$ is zero the diophantine equations for the feedback controller simplify as follows:

Theorem 15

When the measurement noise $n(t) = 0$ the spectral factor D_f is defined by:

$$D_f D_f^* = C\sigma_d C^* + D\sigma_{\ell n} D^* \qquad (2.151)$$

The optimal feedback controller is given by:

$$C_{fb} = \frac{GA_r}{H'A_q} \qquad (2.152)$$

where G, H' (along with F') is the solution having the property:

$$(D_c^* z^{-gl})^{-1} F' \quad \text{strictly proper}$$

of the polynomial equations:

$$D_c^* z^{-gl} G + F'AA_q = z^{-gl} B^* A_r^* B_q^* B_q D_f \qquad (2.153)$$

$$D_c^* z^{-gl} H' - F'BA_r = z^{-gl} A^* A_q^* B_r^* B_r D_f \qquad (2.154)$$

where $gl > 0$ is the smallest integer which makes the equations (2.153)-(2.154) polynomial in d.

The polynomials G and H' in equations (2.153)-(2.154) also

satisfy the diophantine equation:

$$AA_qH' + BA_rG = D_fD_c \tag{2.155}$$

<u>Proof:</u>

To obtain $n(t) = 0$ set $\sigma_n = C_n = 0$, $A_n = 1$. The definition of D_f in equation (2.151) then follows immediately from equation (2.133).

The diophantine equation (2.137) becomes:

$$D_c^*D_f^*z^{-g1}G + FAA_q = z^{-g1}B^*A_r^*B_q^*B_qD_fD_f^*$$

Since D_f^* divides the right-hand-side and the first term on the left side of this equation, it must also divide F. Denoting F by:

$$F = D_f^*F'$$

and cancelling the common factor D_f^* results in equation (2.153).

The diophantine equation (2.138) becomes:

$$D_c^*D_f^*z^{-g1}H - FBA_rA_q = z^{-g1}A^*A_q^*B_r^*B_rD_fA_qD_f^*$$

By a similar reasoning to that used above, D_f^* must again divide F. Since A_q divides the right-hand-side and the second term on the left side of this equation, it must also divide H. Denoting H by:

$$H = A_qH'$$

and cancelling the common factor $A_qD_f^*$ results in equation (2.154).

Using equation (2.136) the feedback controller equation becomes:

$$C_{fb} = \frac{GA_r}{H'A_q}$$

Since $F = D_f^*F'$ the conditions $(D_c^*D_f^*z^{-g1})^{-1}F$ strictly proper, and

$(D_c^* z^{-g1})^{-1} F'$ strictly proper, are clearly equivalent.

Multiplying equation (2.153) by BA_r, equation (2.154) by AA_q and then adding results, using equation (2.132) and cancelling common factors, in equation (2.155).

2.8 OPTIMALITY OF THE IMPLIED DIOPHANTINE EQUATIONS : SDF CASE

In general, calculation of the optimal SDF controller requires
the solution of two couples of polynomial equations : one couple for
the cascade part and another couple for the feedforward part of the
controller. Elimination of the common terms between each of the
coupled equations results in a single equation for each part of the
controller, the _implied_ diophantine equations. Solution of the
original two couples of equations results in the optimal controller,
which shifts both poles and zeros of the closed-loop system to their
desired optimal positions. On the other hand, a controller
calculated using the implied equations ensures only the optimal
positions of the closed-loop poles. The related zeros will not _in_
general be the optimal ones. Solution of the implied equations does
not, therefore, always yield the optimal controller.

The conditions under which solution of the implied equations
does yield the unique optimal controller are derived in this section.
For the cascade part of the controller the analysis is restricted to
the case when the output measurement noise $n(t)$ is zero (Theorem 13).

System description

The SDF system with feedforward is shown in Figure 2.3. As in
Section 2.6 the sub-systems may be represented by use of a least
common denominator polynomial $A = \text{l.c.m}(A_p, A_d, A_x, A_e)$ as:

$$W_p = A^{-1}B \qquad\qquad (2.156)$$

$$W_d = A^{-1}C \qquad\qquad (2.157)$$

$$W_x = A^{-1}D \qquad\qquad (2.158)$$

$$W_r = A^{-1}E \qquad\qquad (2.159)$$

The sub-system W_ℓ is denoted by:

$$W_\ell = A_\ell^{-1} E_\ell \qquad\qquad (2.160)$$

For the sub-system W_n it is assumed that $\sigma_n = C_n = 0$, $A_n = 1$.

Assumptions

1. Each sub-system is free of unstable hidden modes.

2. The plant input-output transfer-function W_p is assumed strictly causal i.e. $<B> = 0$.

3. The disturbance $A^{-1}C$, reference generator $A^{-1}E$, load disturbance $A^{-1}D$, and disturbance generator $A_\ell^{-1}E_\ell$ sub-systems are assumed to be proper rational transfer-functions.

4. It is assumed that the plant data is such that the optimal control problem is solvable, i.e. that conditions (a)-(b) in Theorem 12 hold.

Cost-function

The cost-function which is minimised by the optimal control law is, from equation (2.43):

$$J = \frac{1}{2\pi j} \oint_{|z|=1} \left\{ Q_c \phi_e + R_c \phi_u \right\} \frac{dz}{z} \qquad\qquad (2.161)$$

The error, Q_c, and control, R_c, weighting terms may be expressed as (from equation (2.45)):

$$Q_c = \frac{B_q^* B_q}{A_q^* A_q} \quad , \quad R_c = \frac{B_r^* B_r}{A_r^* A_r} \qquad\qquad (2.162)$$

Assumptions

1. The weighting elements Q_c and R_c are <u>strictly positive</u> on $|d|=1$.

2. A_q, B_q, A_r and B_r are strictly Hurwitz polynomials.

3. The rational functions $A_q^{-1} B_q$ and $A_r^{-1} B_r$ are assumed to be proper.

4. The pairs A_q, A_r A_q, B and A_r, A are each assumed to be coprime.

2.8.1 Optimal cascade controller

<u>Lemma 1</u>

The optimal cascade controller for the system shown in Figure 2.3 with $n(t) = 0$ and the cost-function defined by equation (2.161) is given by:

$$C_c = \frac{G A_r}{H A_q} \qquad (2.163)$$

where G,H (along with F) satisfy the polynomial equations:

$$D_c^* z^{-g1} G + FAA_q = z^{-g1} B A_r^* B_q^* B_q D_f \qquad (2.164)$$

$$D_c^* z^{-g1} H - FBA_r = z^{-g1} A A_q^* B_r^* B_r D_f \qquad (2.165)$$

D_c and D_f satisfy equations (2.105) and (2.119). The diophantine equations must be solved for the minimal solution (G,H,F) with respect to F i.e. the solution such that:

$$(D_c^* z^{-g1})^{-1} F \text{ is strictly proper.}$$

<u>Proof</u>:

Given in Section 2.6.

Lemma 2

The polynomials G and H in equations (2.164)-(2.165) also satisfy the <u>implied</u> cascade diophantine equation:

$$AA_q H + BA_r G = D_f D_c \qquad (2.166)$$

Proof:

Given in Section 2.6.

Optimality of the implied cascade diophantine equation

In general, calculation of the optimal cascade controller polynomials G and H requires solution of the couple of equations (2.164) and (2.165) such that F has minimal degree (condition $(D_c^* z^{-gl})^{-1} F$ strictly proper). The conditions under which the implied cascade diophantine equation (2.166) uniquely determines the optimal cascade controller are now derived.

Preliminaries

Fact 1 (Division Theorem)

For any polynomials E,F there exist <u>unique</u> polynomials Q,R such that:

$$E = QF + R$$

and $F^{-1} R$ is <u>strictly proper</u>

Lemma 3

Let M,N,P be given polynomials. Then the diophantine equation:

$$NY + MX = P$$

possesses a unique solution such that $M^{-1}Y$ is strictly proper, for coprime M,N. Such a solution is said to be minimal with respect to Y.

Proof: (Šebek 1981)

The general solution of the above equation has the form (see Kučera 1979):

$$X = X' + TN$$
$$Y = Y' - TM$$

for a particular solution X', Y' and an arbitrary polynomial T. To prove the Lemma apply Fact 1 to the general solution for Y.

Lemma 4:

Consider the equation:

$$NY + MX = P$$

(where M,N,P are given). If $N^{-1}PM^{-1}$ is strictly proper then: $M^{-1}Y$ is strictly proper iff $N^{-1}X$ is strictly proper.

Proof:

Multiplying the above equation by $N^{-1}M^{-1}$ obtain:

$$M^{-1}Y + N^{-1}X = N^{-1}PM^{-1}$$

Whenever the right-hand-side of this equation is strictly proper (the assumption) then both the left hand side rational transfer-functions are either strictly proper at the same time or neither of them is so.

Corollary 1

The minimal solutions with respect to X and Y of the equation:

$$NY + MX = P$$

may differ in general. If, however, $N^{-1}PM^{-1}$ is strictly proper then the minimal solutions with respect to X and Y are the same.

Main Result

Fact 2

The solution to the optimal SDF control problem is unique and the optimal cascade controller polynomials are given by equations (2.164)-(2.165). The optimal solution is characterised by:

$(D_c^* z^{-gl})^{-1} F$ strictly proper.

Lemma 5

$(D_c^* z^{-gl})^{-1} z^{-gl} B^* A_r^* B_q^* B_q A_q^{-1} A^{-1} D_f$ is a strictly proper rational transfer-function.

Proof:

1. $A^{-1}C$, $A^{-1}D$ and $A^{-1}E$ are proper by definition => $A^{-1}D_f$ is proper (from equation (2.119)).

2. $B_q A_q^{-1}$ is proper by definition.

3. The absolute coefficient of $BA_r B_q$ is zero since the absolute coefficient of B is zero by definition. On the other hand, the absolute coefficient of D_c is non-zero by definition of the spectral factorisation (2.105). As a consequence, $(D_c^* z^{-gl})^{-1} z^{-gl} B^* A_r^* B_q^*$ is a strictly proper rational transfer-function.

Putting together parts (1), (2) and (3) the Lemma results.

Lemma 6

The optimal solution for the cascade controller (characterised by $(D_c^* z^{-g1})^{-1} F$ strictly proper) also has:

$(AA_q)^{-1} G$ strictly proper

Proof:

Consider the strict properness of the transfer-function defined in Lemma 5 and of the transfer-function $(D_c^* z^{-g1})^{-1} F$ (the optimality condition) and then apply Lemma 4 to equation (2.164).

Assumption 1

Let A and B be coprime.

Lemma 7

Let Assumption 1 be satisfied. Then equation (2.166) possesses a unique solution such that:

$(AA_q)^{-1} G$ is strictly proper.

Proof:

By definition, the pairs A_q, A_r A_q, B and A_r, A are coprime. Together with Assumption 1 this means that the pair AA_q, BA_r is coprime. The result then follows immediately by application of Lemma 3.

Theorem 16: <u>Optimality of the implied cascade equation</u>

Let Assumption 1 be satisfied. Then the optimal cascade controller polynomials G and H are determined <u>uniquely</u> by the minimal solution with respect to G of the implied cascade diophantine equation (2.166) i.e. the solution such that:

$(AA_q)^{-1}G$ is <u>strictly proper</u>.

Proof:

By Lemma 6 the optimal solution is characterised by $(AA_q)^{-1}G$ strictly proper. When Assumption 1 holds then, by Lemma 7, the implied equation (2.166) possesses just one solution such that $(AA_q)^{-1}G$ is strictly proper. Such a solution must, therefore, be the unique optimal one.

Discussion

The conditions (1)-(3) in Section 2.8 which the plant must satisfy are physically realistic. The conditions imposed on the cost-function weights are not restrictive and may always be satisfied by suitable selection.

The only further restriction which must be observed for the implied cascade diophantine equation to yield the unique optimal cascade controller is that the plant A and B polynomials must be coprime (Assumption 1). When this condition holds then:

A,B coprime <=> A = A_p

$\qquad$ <=> A_d, A_e and A_x are divisors of A_p

Thus, A,B coprime means that all the poles of the disturbance sub-systems W_d and W_x, and of the reference generator W_r, are poles of the plant input-output transfer-function W_p.

2.8.2 Optimal feedforward controller

Lemma 8

The optimal feedforward controller for the system shown in Figure 2.4 and the cost-function defined by equation (2.161) is given by:

$$C_{ff} = \frac{XA_r D_f - C_{cn}DD_{fd}}{D_{fd}AC_{cd}} \qquad (2.167)$$

where X (along with Z and Y) satisfy the polynomial equations:

$$D_c^* z^{-g^2} X + ZAA_q A_\ell = z^{-g^2} B_r^* A_q^* B_q^* B \, DD_{fd} \qquad (2.168)$$

$$D_c^* z^{-g^2} Y - ZBA_r A_\ell = z^{-g^2} A_q^* A_q^* B_r^* B_r \, DD_{fd} \qquad (2.169)$$

D_c, D_f and D_{fd} satisfy equations (2.105), (2.106) and (2.107) respectively. The diophantine equations must be solved for the minimal solution (X,Y,Z) with respect to Z i.e. the solution such that:

$$(D_c^* z^{-g^2})^{-1} Z \text{ is strictly proper}$$

Proof:

Given in Section 2.6.

Lemma 9

The polynomials X and Y in equations (2.168) and (2.169) also satisfy the implied feedforward diophantine equation:

$$AA_q Y + BA_\ell X = D_c DD_{fd} \qquad (2.170)$$

84

<u>Proof:</u>

Given in Section 2.6.

<u>Optimality of the implied feedforward diophantine equation</u>

<u>Fact 3</u>

The optimal feedforward controller defined by equation (2.167) is unique and the polynomial X is given by equations (2.168)-(2.169). The optimal solution is characterised by:

$(D_c^* z^{-g2})^{-1} Z$ strictly proper

<u>Lemma 10</u>

$(D_c^* z^{-g2})^{-1} z^{-g2} B^* A_r^* B_q^* B_q A_q^{-1} A^{-1} D A_\ell^{-1} D_{fd}$ is a strictly proper rational transfer-function.

<u>Proof:</u>

1. $A_\ell^{-1} E_\ell$ is proper by definition

     => $A_\ell^{-1} D_{fd}$ is proper (from equation (2.107)).

2. $B_q A_q^{-1}$ and $A^{-1}D$ are proper by definition.

3.   By a similar reasoning to that used in the proof of Lemma 5, $(D_c^* z^{-g2})^{-1} z^{-g2} B^* A_r^* B_q^*$ is a strictly proper rational transfer-function.

Putting together parts (1), (2) and (3) the Lemma results.

<u>Lemma 11</u>

The optimal solution for the feedforward controller (characterised by $(D_c^* z^{-g2})^{-1} Z$ strictly proper) also has:

$$(AA_q A_\ell)^{-1} X \quad \text{strictly proper} \tag{2.171}$$

Proof:

Consider the strict properness of the transfer-function defined in Lemma 10 and of the transfer-function $(D_c^* z^{-g2})^{-1} Z$ (the optimality condition) and then apply Lemma 4 to equation (2.168).

Assumption 2

Let A_ℓ be a divisor of both A and D.

Lemma 12

Let Assumption 2 be satisfied. Then the implied feedforward diophantine equation (2.170) becomes:

$$AA_q Y' + BA_r X' = D_c D' D_{fd} \tag{2.172}$$

where:

$$Y \triangleq A_\ell Y' \quad , \quad X \triangleq A_\ell X' \quad , \quad D \triangleq A_\ell D'$$

Proof:

When Assumption 2 holds then, from equations (2.168) and (2.169), A_ℓ must divide both X and Y. Using the above definitions and substituting in equation (2.170) results, after cancellation of the common factor A_ℓ, in equation (2.172).

Lemma 13

Let Assumption 2 be satisfied. Then the optimality condition (2.171) becomes:

$$(AA_q)^{-1} X' \quad \text{strictly proper} \tag{2.173}$$

86

Proof:

When Assumption 2 holds then, from the above definitions, $X = A_\ell X'$. Substituting for X in (2.171) results in (2.173).

Lemma 14

Let Assumptions 1 and 2 be satisfied. Then the implied feedforward equation (2.172) possesses a unique solution such that:

$(AA_q)^{-1}X'$ is strictly proper

Proof:

By definition, the pairs A_q, A_r A_q, B and A_r, A are coprime. Together with Assumption 1 this means that the pair AA_q, BA_r is coprime. The result then follows immediately by application of Lemma 3.

Theorem 17: Optimality of the implied feedforward equation

Let Assumptions 1 and 2 be satisfied. Then the optimal feedforward controller polynomial X is given by $X = A_\ell X'$ where X' is determined uniquely by the minimal solution with respect to X' of the implied feedforward diophantine equation (2.172) i.e. the solution such that:

$(AA_q)^{-1}X'$ is strictly proper

Proof:

By Lemma 13 the optimal solution is characterised by $(AA_q)^{-1}X'$

strictly proper when Assumption 2 holds. When Assumptions 1 and 2 hold then the implied feedforward equation is given by equation (2.172) and possesses just one solution such that $(AA_q)^{-1}X'$ is strictly proper. Such a solution must, therefore, be the unique optimal one.

Discussion

The extra condition required for optimality of the implied feedforward diophantine equation is that A_ℓ, the denominator of the measurable disturbance generator, must divide both A and D. For the unstable measurable disturbance generators of greatest practical interest (such as steps, ramps etc.) this condition also corresponds to one of the optimal control problem solvability conditions (condition (b) in Theorem 12). If, therefore, the measurable disturbance generator is unstable and the optimal control problem is solvable, then the implied feedforward diophantine equation will uniquely determine the optimal feedforward controller whenever A and B are coprime, which is exactly the condition required for optimality of the implied cascade diophantine equation.

2.9 OPTIMALITY OF THE IMPLIED DIOPHANTINE EQUATIONS : 2DF CASE

In general, calculation of the optimal 2DF controller requires the solution of three couples of polynomial equations : one couple for the feedback part, one couple for the reference part and one couple for the feedforward part of the controller. Elimination of the common terms between each of the coupled equations results in a single equation for each part of the controller, the implied diophantine equations. As in the SDF case, solution of the implied equations ensures only the optimal positions of the closed-loop poles. The related zeros will not in general be the optimal ones.

The conditions under which solution of the implied equations does generate the unique optimal controller are derived in this section. For the feedback part of the controller the analysis is restricted to the case when the output measurement noise n(t) is zero (Theorem 15).

System description

The 2DF system with feedforward is shown in Figure 2.4. As in Section 2.7 the sub-systems may be represented by use of a least common denominator polynomial $A = l.c.m(A_p, A_d, A_x)$ as:

$$W_p = A^{-1}B \tag{2.174}$$

$$W_d = A^{-1}C \tag{2.175}$$

$$W_x = A^{-1}D \tag{2.176}$$

The sub-systems W_r and W_ℓ are denoted by:

$$W_r = A_e^{-1}E_r \tag{2.177}$$

$$W_\ell = A_\ell^{-1}E_\ell \tag{2.178}$$

For the sub-system W_n it is assumed that $\sigma_n = C_n = 0$, $A_n = 1$.

Assumptions

1. Each sub-system is free of unstable hidden modes.

2. The plant input-output transfer-function W_p is assumed strictly causal i.e. $<B> = 0$.

3. The disturbance $A^{-1}C$ and $A^{-1}D$ sub-systems, the reference generator $A_e^{-1}E_r$, and the measurable disturbance generator $A_\ell^{-1}E_\ell$ are assumed to be proper rational transfer functions.

4. It is assumed that the plant data is such that the optimal control problem is solvable i.e. that conditions (a)-(c) in Theorem 14 hold.

Cost-function

The cost-function which is minimised by the optimal control law is again given by equation (2.43). The assumptions (1)-(4) on the cost-function weights given in Section 2.8 are assumed to hold.

2.9.1 Optimal feedback controller

Lemma 1

The optimal feedback controller for the system shown in Figure 2.4 with $n(t) = 0$ and the cost-function defined by equation (2.43) is given by:

$$C_{fb} = \frac{GA_r}{HA_q} \tag{2.179}$$

where G,H (along with F) satisfy the polynomial equations:

90

$$D_c^* z^{-g1} G + FAA_q = z^{-g1} B^* A_r^* B_q^* D_f \qquad (2.180)$$

$$D_c^* z^{-g1} H - FBA_r = z^{-g1} A^* A_q^* B_r^* D_f \qquad (2.181)$$

D_c and D_f satisfy equations (2.132) and (2.151). The diophantine equations must be solved for the minimal solution (G,H,F) with respect to F i.e. the solution such that:

$(D_c^* z^{-g1})^{-1} F$ is strictly proper.

Proof

Given in Section 2.7.

Lemma 2

The polynomials G and H in equations (2.180)-(2.181) also satisfy the **implied** feedback diophantine equation:

$$AA_q H + BA_r G = D_f D_c \qquad (2.182)$$

Proof:

Given in Section 2.7.

Optimality of the implied feedback diophantine equation

In general, calculation of the optimal feedback controller polynomials G and H requires solution of the couple of equations (2.180) and (2.181) such that F has minimal degree (condition $(D_c^* z^{-g1})^{-1} F$ strictly proper). The conditions under which the implied feedback diophantine equation (2.182) uniquely determines the optimal feedback controller are now derived.

Assumption 1

Let A and B be coprime.

Theorem 18: Optimality of the implied feedback equation

Let Assumption 1 be satisfied. Then the optimal feedback
controller polynomials G and H are determined <u>uniquely</u> by the minimal
solution with respect to G of the implied feedback diophantine
equation (2.182) i.e the solution such that:

$$(AA_q)^{-1}G \quad \text{is \underline{strictly proper}}$$

Proof

Compare the couples of equations (2.164)-(2.165) and
(2.180)-(2.181), and the implied equations (2.166) and (2.182). The
proof then follows by direct analogy with the derivation of
Theorem 16.

Discussion

In the 2DF case the condition that A,B must be coprime for the
implied feedback diophantine equation to yield the unique optimal
feedback controller may be interpreted as follows:

A,B coprime $<=> A = A_p$

$<=> A_d$ and A_x are divisors of A_p

Thus, A,B coprime means that all the poles of the disturbance
sub-systems W_d and W_x are poles of the plant input-output
transfer-function W_p.

2.9.2 Optimal reference controller

Lemma 3

The optimal reference controller for the system shown in Figure 2.4 and the cost-function defined by equation (2.43) is given by:

$$C_r = \frac{MA_r D_f}{D_m C_{fbd}} \tag{2.183}$$

where M (along with N and Q) satisfy the polynomial equations:

$$D_c^* z^{-g2} M + NA_q A_e = z^{-g2} B^* A_r^* B_q^* B D_m \tag{2.184}$$

$$D_c^* z^{-g2} Q - NBA_r A_{ec}' = z^{-g2} A^* A_q^* B_r^* B_r A' D_m \tag{2.185}$$

D_c, D_f and D_m satisfy equations (2.132), (2.133) and (2.135), respectively. The diophantine equations must be solved for the minimal solution (M,N,Q) with respect to N i.e. the solution such that:

$$(D_c^* z^{-g2})^{-1} N \quad \text{is strictly proper}$$

Proof:

Given in Section 2.7.

Lemma 4

The polynomials M and Q in equations (2.184) and (2.185) also satisfy the **implied** reference diophantine equation:

$$D_e A_q Q + BA_r M = D_c D_m \tag{2.186}$$

Proof:

Given in Section 2.7.

Optimality of the implied reference diophantine equation

Fact 1

The optimal reference controller for the 2DF control problem is unique and the controller polynomial M is given by equations (2.184)-(2.185). The optimal solution is characterised by:

$$(D_c^* z^{-g2})^{-1} N \quad \text{strictly proper}$$

Lemma 5

$$(D_c^* z^{-g2})^{-1} z^{-g2} B^* A_r^* B_q^* B_q A_q^{-1} A_e^{-1} D_m \quad \text{is a strictly}$$

proper rational transfer-function.

Proof:

1. $A_e^{-1} E_r$ is proper by definition

   $\Rightarrow A_e^{-1} D_m$ is proper (from equation (2.135)).

2. $B_q A_q^{-1}$ is proper by definition.

3. By a similar reasoning to that used in the proof of Lemma 5 in Section 2.8.1, $(D_c^* z^{-g2})^{-1} z^{-g2} B^* A_r^* B_q^*$ is a strictly proper rational transfer-function.

   Putting together parts (1), (2) and (3) the Lemma results.

Lemma 6

The optimal solution for the reference controller (characterised by $(D_c^* z^{-g2})^{-1} N$ strictly proper) also has:

$$(A_e A_q)^{-1} M \quad \text{strictly proper}$$

Proof:

Consider the strict properness of the transfer-function defined in Lemma 5 and of the transfer-function $(D_c^* z^{-g2})^{-1} N$ (the optimality condition) and then apply Lemma 4 in Section 2.8.1 to equation (2.184).

Assumption 2

Let A_e be a divisor of A.

Lemma 7

Let Assumption 2 be satisfied. Then the implied reference diophantine equation (2.186) becomes:

$$A_e A_q Q + BA_r M = D_c D_m \qquad (2.187).$$

Proof:

When A_e divides A then, from equation (2.131), $D_e = A_e$. Substituting in equation (2.186) for D_e the Lemma results.

Lemma 8

Let Assumptions 1 and 2 be satisfied. Then the implied reference equation possesses a unique solution such that:

$$(A_e A_q)^{-1} M \quad \text{is strictly proper}$$

Proof:

By assumption, the pairs A_q, A_r and A_q, B are coprime. By Assumption 2, A_e divides A and since by Assumption 1 A,B is a coprime pair so is the pair A_e, B. The pair A_r, A is coprime and since

A_e divides A, so is the pair A_r, A_e. Thus, the pair $A_e A_q$, BA_r is coprime. The result then follows immediately by application of Lemma 3 in Section 2.8.1.

Theorem 19: Optimality of the implied reference equation

Let Assumptions 1 and 2 be satisfied. Then the optimal reference controller polynomial M is determined <u>uniquely</u> by the minimal solution with respect to M of the implied reference diophantine equation (2.187) i.e. the solution such that:

$(A_e A_q)^{-1} M$ is strictly proper

Proof:

By Lemma 6 the optimal solution is characterised by $(A_e A_q)^{-1} M$ strictly proper. When Assumptions 1 and 2 hold then, by Lemma 8, the implied equation (2.187) possesses just one solution such that $(A_e A_q)^{-1} M$ is strictly proper. Such a solution must, therefore, be the unique optimal one.

Discussion

The extra condition required for optimality of the implied reference diophantine equation is that A_e, the reference generator denominator, must divide A (Assumption 2). In the case of the unstable reference generators of greatest practical interest (such as steps, ramps, etc) this condition also corresponds to one of the optimal control problem solvability conditions (condition (b) in

Theorem 14). If, therefore, the reference generator is unstable and the optimal control problem is solvable, then the implied reference diophantine equation will uniquely determine the optimal reference controller whenever A and B are coprime, which is exactly the condition required for optimality of the implied feedback diophantine equation.

2.9.3 Optimal feedforward controller

Lemma 9

The optimal feedforward controller for the system shown in Figure 2.4 and the cost-function defined by equation (2.43) is given by:

$$C_{ff} = \frac{XA_r D_f - C_{fbn} DD_{fd}}{D_{fd} AC_{fbd}} \tag{2.188}$$

where X (along with Z and Y) satisfy the polynomial equations:

$$D_c^* z^{-g3} X + ZAA_q A_\ell = z^{-g3} B_r^* A_q^* B_q^* B DD_{fd} \tag{2.189}$$

$$D_c^* z^{-g3} Y - ZBA_r A_\ell = z^{-g3} A_q^* A_r^* B_r^* B DD_{fd} \tag{2.190}$$

D_c, D_f and D_{fd} satisfy equations (2.132), (2.133) and (2.135), respectively. The diophantine equations must be solved for the minimal solution (X,Y,Z) with respect to Z i.e. the solution such that:

$$(D_c^* z^{-g3})^{-1} Z \quad \text{is strictly proper}$$

Proof:

Given in Section 2.7.

Lemma 10

The polynomials X and Y in equations (2.189) and (2.190) also satisfy the _implied_ feedforward diophantine equation:

$$AA_q Y + BA_r X = D_c DD_{fd} \qquad\qquad (2.191)$$

Proof:

Given in Section 2.7.

Optimality of the implied feedforward diophantine equation

Assumption 3

Let A_ℓ be a divisor of both A and D.

Lemma 11

Let Assumption 3 be satisfied. Then the implied feedforward diophantine equation (2.191) becomes:

$$AA_q Y' + BA_r X' = D_c D' D_{fd} \qquad\qquad (2.192)$$

where:

$$Y \triangleq A_\ell Y' \quad , \quad X \triangleq A_\ell X' \quad , \quad D \triangleq A_\ell D'$$

Proof:

The proof follows by analogy with the proof of Lemma 12 in Section 2.8.2.

Theorem 20 : Optimality of the implied feedforward equation

Let Assumptions 1 and 3 be satisfied. Then the optimal feedforward controller polynomial X is given by $X = A_\ell X'$ where X' is determined _uniquely_ by the minimal solution with respect to X' of the implied feedforward diophantine equation (2.192) i.e. the solution such that:

$$(AA_q)^{-1}X' \quad \text{is strictly proper.}$$

Proof:

The proof follows by direct analogy with the proof of Theorem 17.

Discussion

Again, for the unstable generators of practical importance the condition that A_ℓ must divide A and D corresponds to one of the optimal control problem solvability conditions (condition (c) in Theorem 14). The additional condition that A and B must be coprime is also the condition required for optimality of the implied feedback diophantine equation.

2.10 PROPERTIES AND STRUCTURE OF THE OPTIMAL SOLUTIONS

Some important structural properties of the optimal controller designs may be summarised as follows:

(i)   The dynamic weighting elements in the cost-function allow frequency selective costing to be applied to the tracking error and control input signals. This feature is manifest in the fact that the control weighting denominator A_r is a factor of the numerators of each part of the controller and, when the output disturbance $n(t) = 0$, the error weighting denominator A_q is a factor of the denominators of each part of the controller. Thus, the magnitude of the loop-gain with respect to frequency is directly influenced by the choice of cost weights.

(ii)   The denominator of the output disturbance sub-system (A_n) appears as a zero in the feedback loop. This fact is consistent with the well known transmission-blocking property of zeros (MacFarlane and Karcanias, 1976) and has a natural interpretation since these disturbance modes should not, intuitively, be allowed to propagate through the system.

(iii)   Any Hurwitz poles of the plant input-output transfer function (zeros of A_p) which are not also poles of the disturbance sub-systems (and, in the SDF structure, the reference generator) are cancelled by the controller.

(iv)   In line with the Internal Model Principle of Control
        (Francis and Wonham, 1976) the solvability conditions for
        the optimal control problem demand that any unstable
        reference and disturbance modes must also be modes of the
        plant input-output transfer-function.

(v)   The closed-loop systems for the SDF and 2DF control laws
        are internally stable except in the particular case when
        the plant input-output transfer-function has a pole on
        the unit circle and when the disturbance sub-systems
        (and, in the SDF case, the reference generator) do not.

(vi)   In the SDF controller structure the cascade part of the
        controller is independent of the feedforward part.   In
        the 2DF structure the feedback part of the controller is
        independent of both the reference and feedforward parts.

(vii)   The feedforward part of the controller is causal and
        stable even when the plant is inverse unstable and when
        the delay associated with the plant is longer than the
        delay associated with the measurable disturbance
        sub-system (W_x).   These plant conditions may cause
        serious difficulties in conventional feedforward
        controller designs.

Example 2.1

The objective of this example is to investigate the effect of a constant measurable disturbance (load disturbance) $\ell(t)$ of magnitude 10. For comparison, four cases are considered:

(i)   No feedforward control used, scalar cost weights.

(ii)   No feedforward, dynamic weights with integral action.

(iii)   Conventional feedforward, scalar weights.

(iv)   Optimal feedforward, scalar weights.

The conventional feedforward design is discussed in Section 4.2.3. In this case the feedforward controller is calculated according to:

$$C_{ff} = \frac{D}{B}$$

As discussed in Section 4.2.3, when this design is non-causal (delay in D < delay in B) or unstable (B is unstable) the conventional design is altered to:

$$C_{ff} = \frac{D(1)}{B(1)}$$

In the plant model all measurement noises are assumed to be zero. The plant under consideration is given by

$$W_p = A^{-1}B = \frac{d^2(1+0.5d)}{1-0.95d}$$

$$W_d = A^{-1}C = \frac{1-0.7d}{1-0.95d}$$

$$W_x = A^{-1}D = \frac{d(1-0.75d)}{1-0.95d}$$

The step load disturbance was modelled by:

$$W_\ell = A_\ell^{-1}E_\ell = \frac{1}{1-d}$$

The reference signal was modelled as Gaussian white noise driving the filter:

$$W_r = A_e^{-1} E_r = \frac{1}{1-0.8d}$$

The closed-loop system was simulated over 200 samples, with the load disturbance being applied at sample instant 100.

(i) **No feedforward, scalar cost weights**

When the cost-function weights are chosen as $Q_c = R_c = 0.1$ the feedback and reference controllers may be calculated from Theorems 14 and 15 as:

$$C_{fb} = \frac{0.24}{1.15+0.64d+0.12d^2}$$

$$C_r = \frac{0.63-0.44d}{1.15+0.64d+0.12d^2}$$

The tracking error for this system is plotted in Figure 2.5(a) from which it is seen that the load disturbance is not rejected from the output.

(ii) **No feedforward, integral action**

To obtain integral action the error weighting is chosen as:

$$Q_c = \frac{0.1}{(1-d)^*(1-d)}$$

For this choice of weights the feedback and reference controllers may be calculated as:

$$C_{fb} = \frac{1.53 - 1.23d}{1.35+0.42d-1.12d^2-0.65d^3}$$

$$C_r = \frac{2.54 - 3.32d + 1.08d^2}{1.35+0.42d-1.12d^2-0.65d^3}$$

The tracking error for this system is plotted in Figure 2.5(b). In this case the load disturbance firstly appears on the output and is then rejected by the integral action.

(iii) Conventional feedforward

Since the delay in D is less than the delay in B the conventional feedforward controller is calculated according to:

$$C_{ff} = \frac{D(1)}{B(1)} = 0.167$$

The cost weights were chosen as in (i) to be $Q_c = R_c = 0.1$. The feedback and reference controllers are therefore the same as those in part (i). The tracking error for this system is plotted in Figure 2.5(c). Since the delay in D is less than the delay in B the load disturbance cannot be eliminated initially. The feedforward action does, however, reject the load disturbance in steady state.

(iv) Optimal feedforward

When $Q_c = R_c = 0.1$ the optimal feedforward controller may be calculated from Theorem 14 as:

$$C_{ff} = \frac{1.219 - 0.902d}{1.147 + 0.64d + 0.124d^2}$$

The tracking error for this system is plotted in Figure 2.5(d).

To compare the effectiveness of load disturbance rejection in Parts (ii)-(iv) above the tracking error variance σ_e^2 was approximated in each case by:

104

$$\sigma_e^2 = \frac{1}{200} \sum_{i=1}^{200} e^2(i)$$

In cases (ii)-(iv) the error variance was found to be:

(ii) $\sigma_e^2 = 6.867$

(iii) $\sigma_e^2 = 6.862$

(iv) $\sigma_e^2 = 3.057$

This shows that the integral action and conventional feedforward performance is very similar, while the optimal feedforward results in a tracking error variance approximately half of that in cases (ii) and (iii).

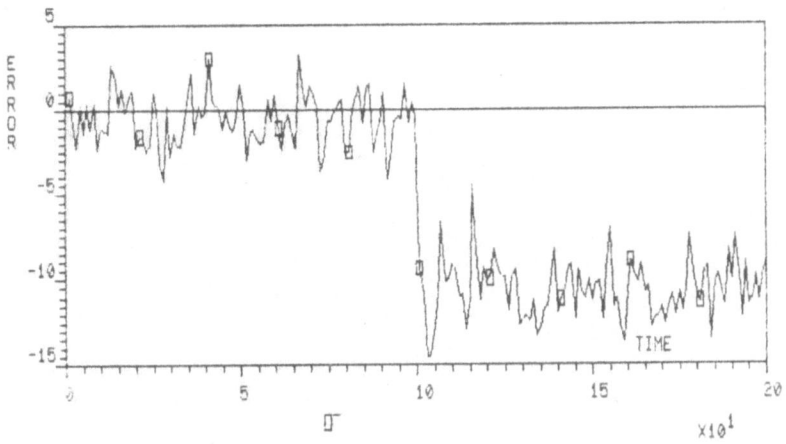

Figure 2.5(a) : No feedforward

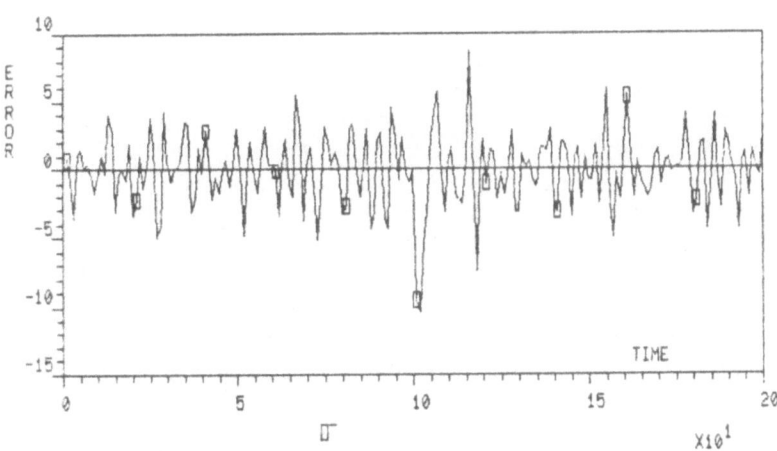

Figure 2.5(b) : Integral Action

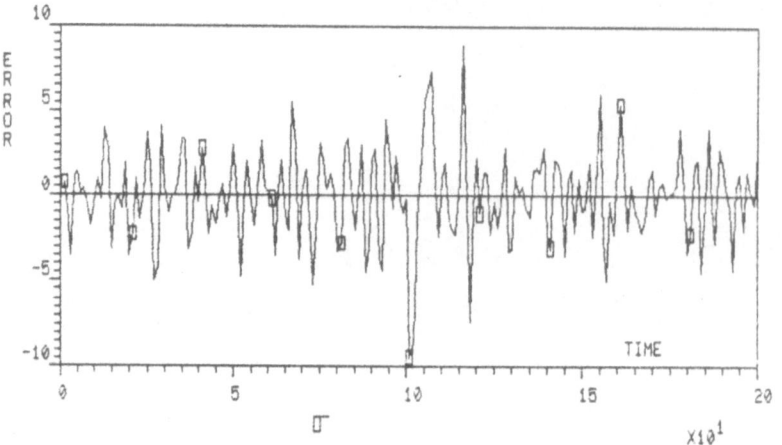

Figure 2.5(c) : Conventional Feedforward

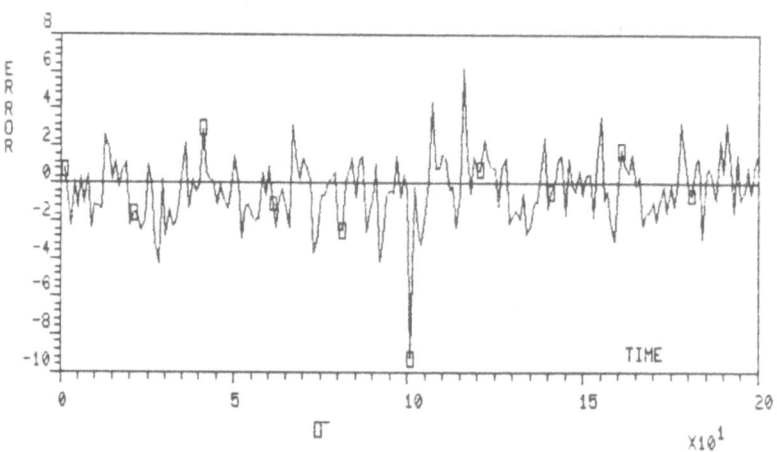

Figure 2.5(d) : Optimal Feedforward

Example 2.2

In this example it is demonstrated that when the plant A and B polynomials have a common factor then the couple of polynomial equations (2.153)-(2.154) must be solved to obtain the unique optimal feedback controller. In addition, the earlier statement that the controller calculated using the implied equation (2.155) when $(A,B) \neq 1$ leads to optimal closed-loop poles but sub-optimal zeros is substantiated.

Consider the following plant:

$$W_p = A^{-1}B = \frac{d(1-0.5d)}{(1-0.5d)(1-0.8d)}$$
$$W_d = A^{-1}C = \frac{1}{(1-0.5d)(1-0.8d)}$$

The polynomials A,B and C may be identified as:

A = (1-0.5d)(1-0.8d)

B = d(1-0.5d)

C = 1

In the optimal control problem to be solved assume that all measurement noises are zero and that the cost-function weights are selected as $A_q = B_q = A_r = B_r = 1$. The spectral factors D_c and D_f may then be calculated using equations (2.132) and (2.151) as:

D_c = (1-0.5d)(1.54-0.52d)

D_f = 1

The solution to equations (2.153)-(2.154) with F minimal degree is:

G = 1.25-0.43d

H = 1.54-0.53d

F = -0.82+1.65d

The unique optimal feedback controller is, from equation (2.152):

$$C_{fb} = H^{-1}G = \frac{1.25-0.43d}{1.54-0.53d} = 0.81$$

If, on the other hand, the implied polynomial equation (2.155) is solved to obtain G and H (after cancelling the common factor (1-0.5d) between A,B and D_c) then the following controller is obtained:

$$C_{fb} = H^{-1}G = \frac{0.71}{1.54} = 0.46$$

Since in both of the above cases the pairs G,H satisfy the implied polynomial equation (2.155) (which is the closed-loop characteristic equation) both controllers give the same (optimal) closed-loop poles as given by the spectral factors. However, equations (2.39) and (2.40) show that the closed-loop zeros depend upon the controller numerator and denominator polynomials G and H. Since in the above example the polynomials G and H calculated using the implied equation are different from the optimal G and H (calculated using the coupled equations) the resulting closed-loop zeros cannot be optimal.

2.11 OPTIMAL REGULATION WITH DISTURBANCE MEASUREMENT FEEDFORWARD –

THE MULTIVARIABLE CASE

The design of optimal regulators for multivariable plants subject to noise disturbances has been intensively studied in recent years. If only the plant output can be measured it is well know that the optimal regulator consists of linear output feedback and can be designed using either time-domain (Kwakernaak and Sivan, 1972) or frequency-domain (Youla et al, 1976b) methods. Alternatively, the optimal multivariable regulator may be designed using the polynomial equation approach developed by Kučera (1979) and extended to the tracking case by Šebek (1983a).

It follows from the preceding sections of this chapter that if, in addition to the plant output, some disturbance can be measured then a two-input controller, utilising both feedback and disturbance measurement feedforward, may be used to improve the controller performance (i.e. to decrease the optimal cost).

The scalar feedback/feedforward regulator solution obtained by Šebek et al (1988) was recently extended to <u>multivariable</u> plants by Hunt and Šebek (1989), and the results of this work are summarised in the following.

Problem Formulation

The multi-input multi-output plant under consideration is governed by the equation:

$$Ay = Bu + C_1 \psi_1 + C_2 \psi_2 \qquad (2.193)$$

where y is the vector output sequence, u is the vector control input sequence and ψ_1 and ψ_2 are two vector noise sequences. A, B, C_1 and C_2 are polynomial matrices in d. The plant is assumed strictly causal, so that $\langle A \rangle$ is invertible while $\langle B \rangle = 0$. The noise component ψ_2 passes through a filter to produce a measured disturbance signal ψ_s i.e.:

$$A_s \psi_s = C_s \psi_2 \qquad (2.194)$$

where A_s and C_s are polynomial matrices in d, with C_s square. The filter $A_s^{-1} C_s$ typically represents measurement dynamics. The general linear controller which operates on the plant output (corrupted by a measurement noise ψ_3) and on the measured disturbance signal ψ_s is described by:

$$Pu = -Q(y + \psi_3) + S\psi_s \qquad (2.195)$$

where P, Q and S are the polynomial matrices to be found, and $\langle P \rangle$ is invertible. The overall system structure is shown in Figure 2.6. Note that in practice the controller must be realised as a single dynamical system having two vector inputs and one vector output (i.e. the control signal u).

All the vector random sources ψ_1, ψ_2 and ψ_3 are mutually independent stationary white noises with intensities σ_1, σ_2 and σ_3, respectively. To avoid the trivial case of $\sigma_2 = 0$ (i.e no measurable disturbance) we assume here, without loss of generality, that $\sigma_2 = I$. σ_1 and σ_3 are real non-negative definite matrices.

The desired optimal controller evolves from minimisation of the cost-function:

$$J = \text{trace}\langle \Omega \Phi_u \rangle + \text{trace}\langle \Sigma \Phi_y \rangle \qquad (2.196)$$

where Φ_u and Φ_y are correlation functions of u and y in steady-state, respectively. Ω and Σ are real non-negative definite weighting matrices. Thus, the design problem is to minimise the cost (2.196) subject to the constraint that the closed-loop system defined by

111

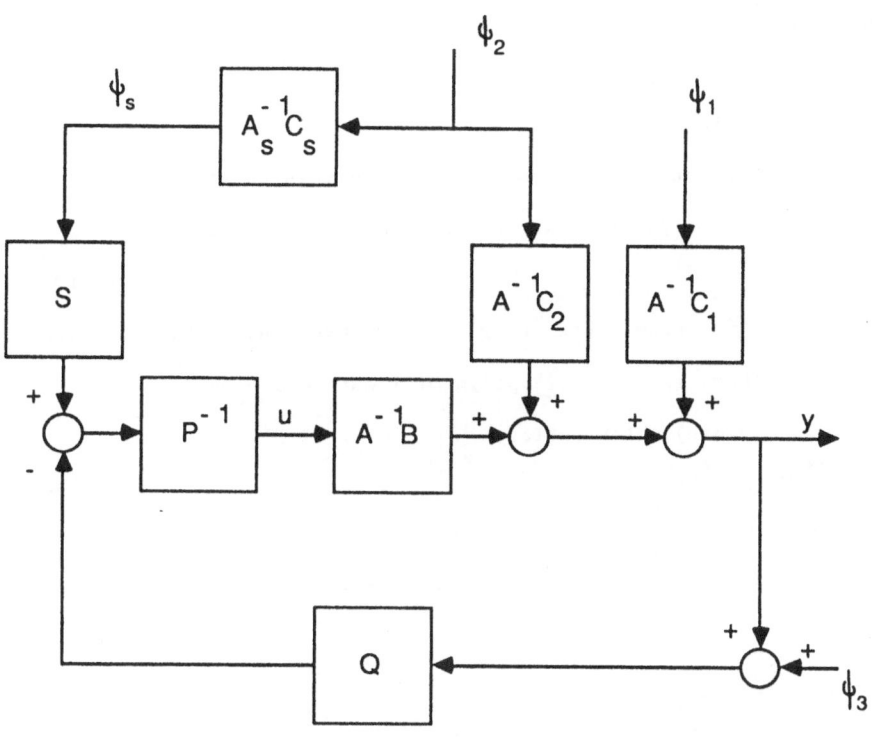

Figure 2.6 : Multivariable regulator with
disturbance measurement
feedforward

equations (2.193)-(2.195) be asymptotically stable.

Problem Solution

The first stage in the design procedure is to find a pair of right-coprime polynomial matrices A_1 and B_1 such that:

$$A^{-1}B = B_1 A_1^{-1} \tag{2.197}$$

For brevity we assume that the given data make the problem regular i.e. that there exist __stable__ polynomial matrices D_c and D_f (the spectral factors) which satisfy:

$$A_1^* \Omega A_1 + B_1^* \Sigma B_1 = D_c^* D_c \tag{2.198}$$

$$A \sigma_3 A^* + C_1 \sigma_1 C_1^* = D_f D_f^* \tag{2.199}$$

Further, the following right-coprime matrix fractions are defined by:

$$D_f^{-1} A = A_d D_{fa}^{-1} \tag{2.200}$$

$$D_f^{-1} B = B_d D_{fb}^{-1} \tag{2.201}$$

$$A^{-1} C_2 = C_a A_c^{-1} \tag{2.202}$$

Finally, we define the right-coprime polynomial matrices B_c, C_b by:

$$BC_b = C_2 B_c \tag{2.203}$$

The main result may now be stated as follows:

Theorem:

The optimal control problem is solvable if and only if:

(i) The greatest common left divisor of A and B is a stable polynomial matrix.

(ii) C_s is a stable polynomial matrix.

The optimal controller polynomial matrices P, Q and S are

obtained from the following left-coprime matrix fraction:

$$\bar{D}_f^{-1}[P,Q,S] = \left[XD_{fb}^{-1}, YD_{fa}^{-1}, (YD_{fa}^{-1}C_a - Z)A_c^{-1}C_s^{-1}A_s \right] \qquad (2.204)$$

Here, X and Y (along with V) is the solution of the equations:

$$D_c^* X + V^* B_d = A_1^* \Omega D_{fb} \qquad (2.205a)$$

$$D_c^* Y - V^* A_d = B_1^* \Sigma D_{fa} \qquad (2.205b)$$

such that $\langle V \rangle = 0$.

The polynomial matrix Z (along with U and W) is the solution of the equations:

$$D_c^* U + W^* B_c = A_1^* \Omega C_b \qquad (2.206a)$$

$$D_c^* Z - W^* A_c = B_1^* \Sigma C_a \qquad (2.206b)$$

such that $\langle W \rangle = 0$.

Proof:

We define six rational matrices p, q, s, t, p_1 and q_1 by:

$$p^{-1}[q,s] = P^{-1}[Q,S] \qquad (2.207)$$

and

$$\begin{bmatrix} A & B \\ q & -p \end{bmatrix} \begin{bmatrix} P_1 & B_1 \\ q_1 & -A_1 \end{bmatrix} = I \qquad (2.208)$$

$$t = qA^{-1}C_2 - sA_s^{-1}C_s \qquad (2.209)$$

Using equations (2.193)-(2.195) the vector control input and output sequences may be expressed as:

$$u = -A_1 q A^{-1} C_1 \psi_1 - A_1 t \psi_2 - A_1 q \psi_3 \qquad (2.210)$$

$$y = (I - B_1 q)A^{-1}C_1 \psi_1 - (B_1 t - A^{-1}C_2)\psi_2 - B_1 q \psi_3 \qquad (2.211)$$

The corresponding correlation functions are:

$$\Phi_u = A_1 q A^{-1} C_1 \sigma_1 C_1^* A^{*-1} q^* A_1^* + A_1 t \sigma_2 t^* A_1^* + A_1 q \sigma_3 q^* A_1^*$$

$$= A_1 q A^{-1} D_f D_f^* A^{-1^*} q^* A_1^* + A_1 t \sigma_2 t^* A_1^* \tag{2.212}$$

$$\Phi_y = (I - B_1 q) A^{-1} D_f D_f^* A^{-1^*} (I - q^* B_1^*) + B_1 q \sigma_3 + \sigma_3 q^* B_1^*$$
$$- \sigma_3 + (B_1 t - A^{-1} C_2) \sigma_2 (B_1 t - A^{-1} C_2)^* \tag{2.213}$$

We now substitute (2.212)-(2.213) in (2.196). On employing (2.198)-(2.199) and completing the squares the cost may be expressed in the form:

$$J = \text{trace} \langle G_q^* G_q \rangle + \text{trace} \langle H_q \rangle + \text{trace} \langle G_t^* G_t \rangle + \text{trace} \langle H_t \rangle$$
$$- \text{trace} \langle \Sigma \sigma_3 \rangle \tag{2.214}$$

where,

$$G_q = (D_c q - D_c^{-1^*} B_1^* \Sigma) A^{-1} D_f \tag{2.215}$$

$$H_q = D_f^* A^{-1^*} (\Sigma - \Sigma B_1 D_c^{-1} D_c^{-1^*} B_1^* \Sigma) A^{-1} D_f \tag{2.216}$$

$$G_t = (D_c t - D_c^{-1^*} B_1^* \Sigma A^{-1} C_2) \tag{2.217}$$

$$H_t = C_2^* A^{-1^*} (\Sigma - \Sigma B_1 D_c^{-1} D_c^{-1^*} B_1^* \Sigma) A^{-1} C_2 \tag{2.218}$$

Now, notice that the first term in (2.214) is related to the feedback term q and does not depend on the feedforward term s. Conversely, the third term in (2.214) is related to the feedforward term s and does not depend on the feedback term q. The remaining terms in (2.214) are not affected by the controller at all. To achieve the minimum cost we therefore minimise the first and third terms in (2.214) independently by suitable choice of q and s. Minimisation of the first term is known (Kučera, 1979) to be accomplished by setting:

$$p = D_c^{-1} X D_{fb}^{-1} \quad , \quad q = D_c^{-1} Y D_{fa}^{-1} \tag{2.219}$$

where X and Y are given by (2.205). Consequently, due to the definitions of D_c, D_{fa} and D_{fb}, p and q are stable rational matrices.

To minimise the third term in (2.214) we proceed analogously : using equation (2.206b) G_t is decomposed as:

$$G_t = (D_c t - ZA_c^{-1}) + D_c^{-1} W^*$$ (2.220)

Since the equation (2.206) is solved subject to $\langle W \rangle = 0$, the best which can be done to minimise $\langle G_t^* G_t \rangle$ is to set the term in brackets in (2.220) to zero. This calls for:

$$t = D_c^{-1} Z A_c^{-1}$$ (2.221)

or,

$$s = D_c^{-1}(Y D_{fa}^{-1} C_a - Z) A_c^{-1} C_s^{-1} A_s$$ (2.222)

where use has been made of (2.209), (2.202) and (2.219). Equation (2.204) then follows from (2.219), (2.222) and (2.207), (2.208).

Now we must show that, similarly to p and q, s defined by (2.222) is also a stable rational matrix. To this end, post-multiply (2.205a) by $D_{fb}^{-1} C_b$ and (2.205b) by $D_{fa}^{-1} C_a$. By comparing the left-hand sides of the resulting equations, and using the identities (2.200)–(2.203), we obtain:

$$D_c^* \left[X D_{fb}^{-1} C_b - U, \ Y D_{fa}^{-1} C_a - Z \right] = (V^* D_f^{-1} C_2 - W^*) \left[-B_c, \ A_c \right]$$ (2.223)

By the definitions (2.202)–(2.203) and by condition (i) all the invariant polynomials of $\left[-B_c, \ A_c \right]$ are stable. It follows that D_c^* (unstable) divides the right-hand-side of (2.223) and A_c divides the left-hand-side. It follows, therefore, that s defined by (2.222) is stable.

To complete the proof we must still justify the solvability conditions (i) and (ii). Clearly, the cost is finite iff all the rational matrices in (2.210) and (2.211) are stable. It follows from equations (2.210) and (2.211) that this is the case iff p, q, and s

116

are stable and both conditions (i) and (ii) hold. Moreover, condition (i) implies the existence and uniqueness of the solution to equations (2.205) and (2.206). See Kučera (1979) for a proof of this assertion.

Finally, using the theorem on the stability of multivariable feedback systems given by Kučera (1979), stable p, q, and s result in an asymptotically stable closed-loop system.

PART TWO

SELF-TUNING CONTROL

CHAPTER THREE

INTRODUCTION TO SELF-TUNING CONTROL

Summary

This chapter provides an introduction to Part 2 of the thesis.
The fundamental reasons for the use of feedback control (as opposed
to open-loop control) are reviewed in Section 3.1.   The ubiquitous
PI control law is discussed in Section 3.2, and its associated
problems then lead to the introduction of analytical design, adaptive
control and identification in Sections 3.3 and 3.4.   A brief history
of adaptive control is given in Section 3.5 and the three most widely
used approaches to adaptive control (gain scheduling, model reference
adaptive control and self-tuning control) are discussed in Section
3.6.   Previous approaches to self-tuning control are briefly reviewed
in Section 3.7.   Finally, the contributions made in Chapter 4 in the
polynomial equation approach to self-tuning control are outlined.

3.1 FEEDBACK CONTROL

The overall purpose of a control system is depicted, in a very
general sense, in Figure 3.1 : the process responses are required to
be related in a specified way to the system inputs.   To achieve the
desired response the process is driven by a set of actuating signals
which are generated by the system controller.   The control synthesis
task is to design a controller which, from measurements of the system
inputs, will generate the required actuating signals.

In the simplest problem where the system has only one input and
one response signal the design of an appropriate controller may, at
first sight, appear to be very straightforward.   Suppose the process

is described by a transfer-function W(s) and that the desired overall transfer-function from input to output is T(s). Any desired relation T(s) between input and response may be realised by an open loop configuration as shown in Figure 3.2. The controller transfer-function C(s) is

$$C(s) = \frac{T(s)}{W(s)} \tag{3.1}$$

In the open-loop configuration the overall transfer-function T(s) is obtained by cancellation of the process dynamics since the controller contains the inverse of the process transfer-function. Clearly, however, the success of the open-loop solution depends upon accurate a priori knowledge of the process characteristics. Such an approach fails in the following circumstances:

(i)   When W(s) is not accurately known in advance of controller design.

(ii)   When W(s) varies during normal system operation.

(iii) When the process is subject to unknown disturbances which corrupt the responses.

In addition, the cancellation of any unstable poles of W(s) will create an unstable hidden mode in the forward path. To circumvent the difficulties posed by the above three factors, which characterise almost all real design problems, a feedback configuration is normally used. A typical feedback control system is shown in Figure 3.3 where the disturbances affecting the process have been included.

The equation relating the command input r and disturbance signal n to the process output y may be found by straightforward analysis as

121

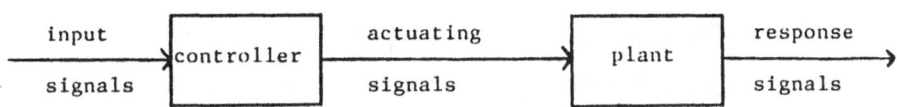

Figure 3.1 : The general control
Problem

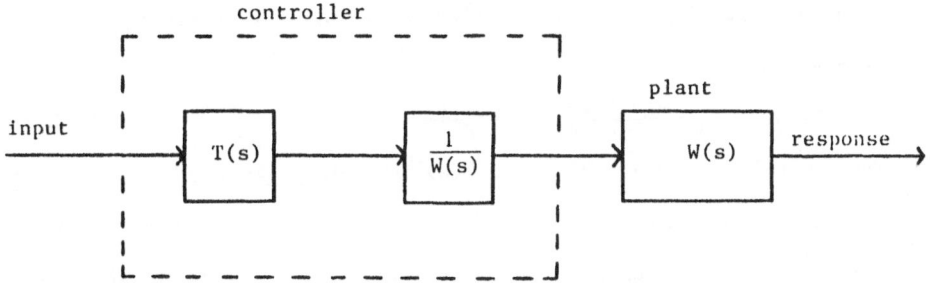

Figure 3.2 : Open-loop Control

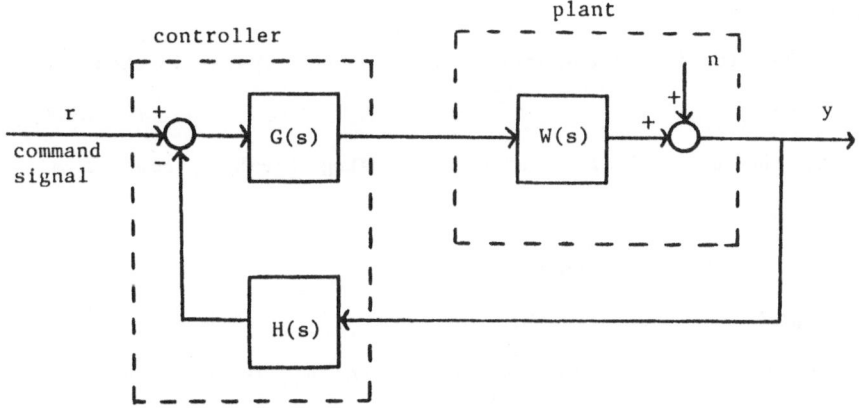

Figure 3.3 : Feedback Control

$$y = \frac{W(s)G(s)}{1 + W(s)G(s)H(s)} \ r + \frac{1}{1 + W(s)G(s)H(s)} \ n \qquad (3.2)$$

The closed-loop **sensitivity function** S(s) is defined as

$$S(s) = \frac{1}{1 + W(s)G(s)H(s)} \qquad (3.3)$$

If at all frequencies of interest (i.e all frequencies where the desired system response signal contains significant energy) G(s) is made sufficiently large so that

$$|W(j\omega)G(j\omega)H(j\omega)| \gg 1 \qquad (3.4)$$

then the overall transfer-function from input to response at these frequencies is

$$T(s) = \frac{W(s)G(s)}{1 + W(s)G(s)H(s)} \ \simeq \ \frac{1}{H(s)} \qquad (3.5)$$

The overall transfer-function T(s) is **independent** of the process dynamics W(s) and is therefore unaffected by uncertainty or variations in the process characteristics. In addition, when (3.4) holds the sensitivity function, which is equivalent to the transfer-function between the disturbance signal and the process response, becomes

$$S(s) \simeq 0 \qquad (3.6)$$

resulting in the elimination of the effect of the disturbance from the process response.

The closed-loop equation of the feedback system is then

$$y \simeq \frac{1}{H(s)} \ r \qquad (3.7)$$

Thus, the two primary reasons for using **feedback** control are to reduce the effects of

    (i)   Parameter uncertainty

    (ii)   Unknown disturbances

where parameter uncertainty is taken to include both initial ignorance of W(s) and subsequent variations in W(s).

3.2 CONVENTIONAL CONTROL

Probably the most widely used form of feedback control law in industry today is the Proportional Integral (PI) controller. The basic form of the PI controller is illustrated in Figure 3.4. The control law which generates the actuating signal u(t) is described by

$$u(t) = K(e(t) + \frac{1}{T_i} \int^t e(m)dm) \qquad (3.8)$$

where the proportional gain K and the integral time-constant T_i are the control design parameters which must be selected by the control system designer.

The selection of appropriate values of K and T_i to achieve a desired response is known as the tuning problem. When tuning a control loop the engineer typically uses one of two techniques:

(i)   The control loop is opened and the process input u(t) is manipulated manually. From the process response the appropriate values of K and T_i can be obtained using some heuristic rule (such as the Ziegler-Nichols (1942) method).

(ii)   A trial-and-error approach can be adopted. The engineer makes an initial guess of the values of K and T_i and, based on observation of the closed-loop response obtained, subsequently changes these values to achieve the desired response.

It has been estimated by some sources that around 80% of existing process control loops are based on the PI control law and have been tuned by one of the methods outlined above (Deshpande and Ash 1980). Typically, after a loop has been tuned the values of the controller coefficients remain unchanged over the operational lifetime of the controller. This situation is not surprising since modern process plants have a very large number of loops, many of which have time constants of the order of minutes or even hours. The tuning of loops using the trial-and-error or heuristic methods outlined above can therefore be a very time consuming procedure and it is difficult to obtain a set of controller coefficients which in some way can be regarded as being the 'best'.

Although conventional controllers (such as PI) are widely regarded as giving adequate control performance, the tuning difficulty means that the accepted performance is almost always inferior to that which is ultimately possible. This fact is becoming more important as tighter control on new and existing loops is demanded to ensure that plants are operated as efficiently as possible. Small improvements in control performance can result in large economic benefits due, for example, to savings in raw materials and energy. A further important feature of conventional control designs is the simplicity of the control laws employed. While it is true that many control loops in process control applications can be reasonably approximated by a well-damped second order transfer-function, more 'advanced' applications and a few critical process control loops exhibit dynamics which render a simple control law such as PI inadequate and demand a more complex design. Successful implementation of a PI controller is made particularly difficult in the presence of any of the following conditions:

(i)   The process may contain significant dead-time i.e. it may contain an appreciable time-delay between a given input to the process and the resulting response.

(ii) The process may be of high order.

(iii) The process may be open-loop unstable, poorly damped, or non-minimum phase.

A final factor in the consideration of conventional control system design is the variation of process dynamics during system operation. A controller which is well tuned initially may exhibit unsatisfactory performance should the dynamic characteristics of the controlled process change. Such dynamic changes may be due to several causes, for example:

(i)   Changes in environmental conditions.

(ii) Ageing of system components.

(iii) Non-linearities, where the process gain varies with the operating point.

An overview of conventional process control system design is given by Shinskey (1979).

3.3 ANALYTICAL DESIGN AND ADAPTIVE CONTROL

In distinct contrast to the conventional design methods described above are the range of analytical design techniques discussed in Chapter 1 which have been developed by control theorists during the past forty years. By analytical design is meant the application of mathematical techniques to idealised models which represent the physical process to be controlled : given a process model the design algorithm produces once-and-for-all the controller which meets the demanded performance specification.

Clearly, the concept of analytical design is highly idealised and, assuming an accurate model of the process is available, depends on two factors:

(i)   The formulation of a 'sensible' specification of control performance.

(ii) The mathematical tractability of the design problem as defined by the performance specification and process model.

The analytical design techniques provide a sharp contrast to the trial-and-error methods since they proceed from the problem specification directly to the final controller design without the need for subjective analysis.   Further, the methods place no restriction on the complexity of the controller and most can therefore cope with processes which have complex dynamics.   For these reasons the analytical design techniques overcome the first two of the drawbacks of conventional controllers listed above.

Implicit in the discussion of the analytical design techniques is the assumption that a model of the process is available.   The success of any such design in achieving the specified performance objectives depends directly upon the accuracy with which the process dynamics are known :   only if the model did <u>exactly</u> represent the actual process dynamics would the performance specification be <u>exactly</u> met.   The application of any analytical design can only follow an evaluation of the process dynamics.

Evaluation of process dynamics is known as the <u>identification</u>

problem. Techniques for identifying the process dynamics are discussed briefly in the following section. For the purposes of controller design it is most useful to identify process dynamics in transfer-function form.

Assuming that the initial identification of the process leads to an accurate model, the problem of initial ignorance of the process dynamics can be overcome and a suitable controller can then be designed. However, the problem of subsequent variation in process dynamics remains. As with conventional control, variations in process dynamics can lead to a deterioration in control performance. The fact that almost all real physical processes display some kind of time variation in their dynamics has led to the phenomenal growth of interest in the concept of adaptive control : control in which the automatic and continual identification of process dynamics is used as a basis for the automatic and continuing re-design of the controller.

The general concept of adaptive control is encapsulated in Figure 3.5. Compared to the non-adaptive control scheme in Figure 3.1 the adaptive controller consists of two additional elements:

(i)   Process identification. The identifier, using measurements of various process signals, determines the dynamic characteristics of the process.

(ii)   Controller adjustment. The controller coefficients are continually adjusted in sympathy with any measured variation in process dynamics.

Thus, the ideal adaptive controller can overcome the problems posed by parameter variation and continually meet the demanded

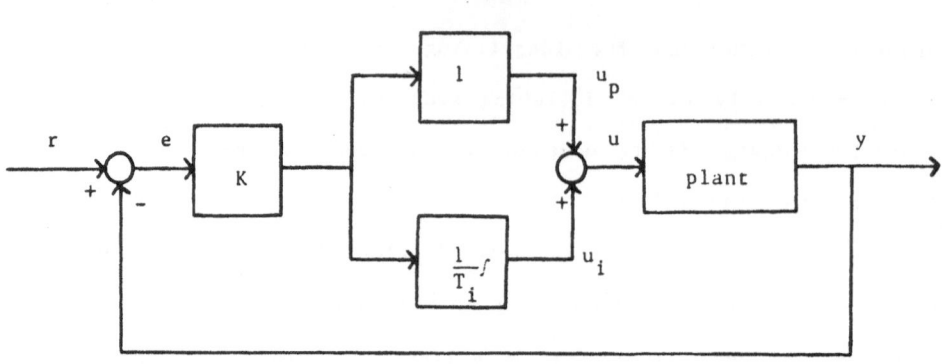

Figure 3.4 : PI Controller

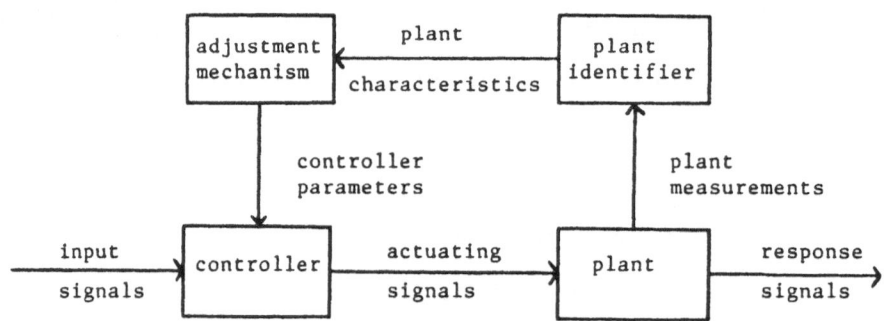

Figure 3.5 : The General Adaptive
            Controller

performance specification. In Section 3.1, however, reduction of the effect of parameter variation was cited as a principal reason for the use of feedback control. Why, then, is adaptation required in addition to feedback?

Unfortunately, the high gain which ensures that condition (3.4) is satisfied at frequencies of interest also tends to drive the closed-loop system into instability at other critical frequencies. In addition, this high gain will tend to accentuate any measurement noise which is present. A limit exists in the allowable closed-loop gain, which essentially amounts to a trade-off between performance and stability. Although the effect of parameter variation is greatly reduced by feedback, the degree to which the overall design objectives are achieved still depends critically upon the level of available knowledge about the process dynamics.

3.4 SYSTEM IDENTIFICATION

Most areas of engineering and scientific inquiry are concerned with the study or manipulation of dynamical systems (systems where the present output depends not only on its present input, but on its past history i.e. systems having memory). Central to the study of dynamical systems is the concept of a system model; a model provides a convenient means of summarising knowledge about the system's properties and behaviour.

System models

A system model can assume many forms, for example:

(i)   <u>Mental</u> or <u>intuitive</u> models where the knowledge about the
        system is held in the mind of the person interacting with
        that system. For instance, a driver will generally build
        up an intuitive feel for the way in which a vehicle
        responds to the various inputs (accelerator position or
        steering wheel angle, for example) he or she applies to
        the system.

(ii)   <u>Graphical</u> models where the system properties are summarised
        in a graph or table; in process control a graph of the
        non-linear characteristics of a valve is frequently
        used.

(iii)   <u>Mathematical</u> models where the system properties are
        summarised by the mathematical relationship between
        system variables; Newton's second law provides a
        universal model which states that the acceleration of a
        body is directly proportional to the force acting on it.

These few examples illustrate firstly the general necessity of models
as an aid to the understanding of dynamical systems and secondly
their role in facilitating interaction with those systems. By far
the most important class of models, however, are mathematical models.
The following list of applications, which is by no means exhaustive,
helps to clarify the importance of and need for mathematical models:

(i)   Throughout science mathematical models are used to quantify
        the gross features of system behaviour.   The model is
        then used to infer the more general system properties and
        to examine the wider implications which result.

(ii)   A model allows the future behaviour of a system to be
        predicted.   Prediction models find application in such

areas as economics and control.

(iii)   The analytical control system design techniques outlined
in a previous section require that a model of the
controlled process be available.

(iv)   Models can be used for fault diagnosis. When the
measured system behaviour is seen to differ significantly
from the model behaviour this may indicate a fault
condition in the system.

(v)   Models can be used for system simulation and operator
training.   Examples are simulators for aircraft and
nuclear power stations.   Such simulators also allow
unusual or potentially hazardous situations to be
investigated.

There is one class of mathematical models which has received far more
attention than any other, namely linear, lumped, time-invariant
models. Briefly, this class of models is characterised as follows:

(i)   Linear models: If the response of a system to an input
$u_1(t)$ is $y_1(t)$ and its response to $u_2(t)$ is $y_2(t)$, it is
linear if its response to $\alpha u_1(t) + \beta u_2(t)$ is $\alpha y_1(t) +
\beta y_2(t)$, where α and β are real constants.

(ii)   Lumped models: If a system's variables are functions of
time only and have no spatial dependence, then the system
is lumped. Otherwise, the system is distributed.

(iii)   Time-invariant models: A dynamical system is time-
invariant if its input-output relations do not vary with
time i.e. if the response of a time-invariant system to
an input u(t) is y(t), then the response to the delayed

input $u(t-\tau)$ is $y(t-\tau)$.

The predominant factors in the popularity of linear, lumped, time-invariant models are their simplicity and amenability to analytical manipulation. Within this class of model there are several further possibilities regarding the precise nature of the model. Some of the most important distinctions are:

(i)   Internal or external models. Internal models (such as state-space models) of a dynamical system describe all the internal couplings between the system variables. External models, or input-output models, describe only the relationship between the system input and output.

(ii)   Time-domain or frequency-domain models. System models may be represented in the time-domain using differential or difference equations. Alternatively, the system may be described by a transfer-function in the frequency domain using either Laplace- or z-transform techniques.

(iii) Continuous or discrete models. Most real systems are by their very nature continuous-time. In very many cases, however, observation of a system is performed using a computer, so that the available data is discrete-time. In addition, the growing popularity of digital control techniques calls for discrete-time models.

(iv)   Deterministic or stochastic models. If the response of a system to a given input is certain, then the system is deterministic. Frequently, however, system responses are subject to stochastic, or random variations due, for example, to noise disturbances.

The majority of analytical control design techniques used in adaptive control are based on models which are linear, lumped and time-invariant. Invariably, the particular form of model used is an external transfer-function in discrete-time. In the subsequent discussion, therefore, this type of model is assumed.

Although the control design techniques and identification methods are based on the assumption that the controlled process is time-invariant, the main motivation for the use of adaptive control is the requirement to maintain a specified control performance in the face of variations in process dynamics. Such systems can still be considered time-invariant if the dynamics vary slowly in comparison with the response time of the overall system. Satisfactory performance is then obtained by continually updating the system model upon which the control design is based. Techniques which allow variations in process dynamics to be tracked are described in Chapter 4. These techniques amount to small modifications of the basic identification methods.

How to construct a system model

There are two basic ways in which a mathematical model of a system can be constructed: from prior knowledge about the system or by analysis of experimental data obtained from the system. These approaches are known, respectively, as modelling and identification:

(i) Modelling: The internal mechanisms which shape the behaviour of a system can be investigated. By direct analysis of the physical laws governing the system, a mathematical model can be constructed.

(ii)   Identification: A mathematical model can be constructed by performing experiments to obtain data from the system. Various techniques can then be used to determine the model which best fits the measured data.

In practice the distinction between modelling and identification is not quite so clear-cut and most models are built using a mixture of the two techniques.   In the modelling approach it is frequently impossible to build a complete model of the system due to a lack of total knowledge about the physical laws governing the system's behaviour.   Thus, mathematical modelling is usually combined with experimentation.   A further feature of the modelling approach is that it may be difficult and time-consuming.

In identification, on the other hand, it is clearly desirable to plan experimental trials using as much prior knowledge about the system as possible.   Identification provides the foundation upon which the majority of adaptive control techniques are built.

Identification

The techniques of constructing a mathematical model using measured data consist of the following steps:

(i)   Experiment design: It is necessary to ensure that the input to a system during an identification experiment is sufficiently 'rich'. This ensures that all modes of the system are excited so that the measurements contain relevant information about the system dynamics. This issue is of particular importance when a process is identified while under closed-loop control since the process input cannot then be freely chosen.

(ii) <u>Choice of model structure</u>: The exact parameterisation of the model to be identified must be chosen before experimentation begins. Often, the model is simply chosen to be linear and of finite order.

(iii) <u>Parameter estimation</u>: Having decided upon the model structure, the parameters, or coefficients, of that model are obtained by processing the measured data. The most common form of parameter estimation methods are formulated as optimisation problems where the best model is selected as the one that best fits the measured data as judged by a specified criterion.

(iv) <u>Model validation</u>: After a model has been obtained from an identification experiment, it must be checked to ensure that it is a credible representation of the actual system. Any inadequacies which become apparent may require alterations to the model structure or experimental conditions. In practice, therefore, system identification is an iterative procedure.

The many identification techniques which are available can be split into two broad classes:

(i) <u>Off-line</u> (batch) methods. In off-line identification a batch of data is collected by taking measurements during an experimental run. After the experiment is complete the data is processed to produce a model.

(ii) <u>On-line</u> methods. In on-line identification a <u>recursive</u> algorithm is used to update a model at each time instant as new data becomes available. Such <u>recursive</u> <u>identification</u> methods are used in adaptive control and in other real-time applications where the process dynamics must continually be monitored.

In applications where the possibility of performing off-line identification exists the estimates obtained are usually of higher precision and are more reliable. The storage requirements for off-line algorithms are, however, far greater than those for on-line methods. In the recursive on-line methods only the most recent data must be retained and old data can be discarded. The methods of recursive system identification are the subject of the recent books by Ljung and Söderström (1983) and Norton (1986), and the survey paper by Hunt (1986).

Methods of identification

Some of the classic methods of identification are based upon non-parametric system models. An example of such a model is a system's impulse response which is specified directly by each value of it's argument. Other classic methods obtain the system transfer-function by frequency-response or transient-response analysis. In the presence of noise, correlation techniques have frequently been used.

An alternative to the classic methods are the 'best-fit' methods in which a criterion function is introduced to give a measure of how well a model fits the experimental data. The most common of such methods is the <u>least-squares</u> technique. In this technique the model parameters are selected in such a way that the sum of the squared errors between the model output and the measured system output is minimised; if the model parameters are denoted by the vector θ and $\hat{y}(\theta)$ represents the model output, which ideally should equal the actual output y, then the least-squares method attempts to find the model parameters such that the criterion

$$J(\theta) = \frac{1}{2} \sum_{i=1}^{N} (y - \hat{y}(\theta))^2 \qquad (3.9)$$

is minimised, where i = 1, 2...N represents the discrete instants of time over which the identification experiment is performed.   The identification methods described in Chapter 4 are based on the principle of least-squares.

3.5 ADAPTIVE CONTROL - A BRIEF HISTORY

Although the idea of a control system which has the ability to continuously adapt to changing process conditions has a strong intuitive appeal, the initial interest in adaptive control arose, like many other major developments in control theory, from the need to solve an important engineering problem.   Research in adaptive control first became very active in the early 1950's in connection with the design of control systems for high performance aircraft (Gregory 1959, Mishkin and Braun 1961).   The performance characteristics of these aircraft varied significantly in flight due to the wide operational range of speed and altitude.   It was found that a normal fixed parameter controller could only be matched to a single flight condition and that at other conditions within the flight envelope the fixed controller would give unsatisfactory performance.   The need to develop a more sophisticated controller which could adapt to changing dynamic characteristics generated a great deal of enthusiasm and effort throughout the fifties. Early surveys in this area are given by Aseltine et al (1958), Stromer (1959), Jacobs (1961) and Truxal (1964). Applications at this time were, however, largely unsuccessful. The adaptive concept seemed to be a natural way to deal with the parameter variation problem but the lack of initial success was the result of two major factors:

(i)   Existing hardware was not sufficiently advanced to deal with the additional complexity of the adaptive controllers.

(ii)   A comprehensive theory of the main aspects of adaptive control was not available.

Enthusiasm in adaptive control was lessened to some extent in the sixties. In addition to the above problems, this situation was brought about by the emergence of a major new technological challenge: in the USA and USSR enormous resources were channelled into the research and development of control systems for the guidance and tracking of space vehicles. Included in the rapid progress made during this time were many contributions which proved to be of great importance for the development of adaptive control : major advances were made in the theory of stochastic control, in system identification, and in estimation theory.

These developments led in the early seventies to a renewed interest in adaptive control and three factors were to play a key role in its success:

(i)   The epoch-making progress in microelectronics made it possible to implement the new generation of adaptive control algorithms easily and cheaply.

(ii)   Some of the basic theoretical issues in adaptive control
        were addressed and solutions began to appear.

(iii)   A catalogue of successful industrial applications began
        to emerge as understanding of the fundamental
        implementation problems increased.

With these developments came the widespread belief amongst
control researchers that the early promise of adaptive control could
eventually be realised. The success of practical trials also
indicated that adaptive control was valuable in applications other
than the advanced flight control systems which prompted the initial
interest in the area. Most of the successful applications were in
fact in industrial process control problems. The eighties have seen
a continuation of the vigorous development of adaptive control. A
measure of the importance of the techniques is the emergence in
recent years of commercially available adaptive controllers, a trend
which continues to grow rapidly. This trend has been accompanied by
the awakening of a strong interest in adaptive control amongst
practicing industrial control engineers. The feedback received from
industry will hopefully be a major contribution to the continued
maturing of the basic techniques. Surveys of the theory and
application of the various adaptive control techniques are given by
Unbehauen (1980), Narendra and Monopoli (1980), Harris and Billings
(1981), Åström (1983a) and Warwick (1988).

3.6   THE METHODS OF ADAPTIVE CONTROL

There have been many attempts to derive techniques which change the controller parameters in response to changes in process dynamics. The three most widely studied classes of adaptive control algorithms are gain scheduling controllers, model-reference adaptive controllers and self-tuning controllers.

The generic single-input single-output adaptive feedback control system is shown in Figure 3.6. The scheme consists of a normal feedback control loop with process and controller. The controller, however, has parameters which are adjusted by an outer loop consisting of process identifier and adjustment mechanism. The three classes of adaptive algorithm differ only in the way in which the controller parameters are adjusted.

Gain scheduling

In some applications where the dynamics of the controlled process are known to exhibit time variation it is possible to find a process variable which changes in sympathy with changing dynamics. If such a variable can be measured, then changes in process dynamics can be inferred. It is then possible to derive a schedule of controller settings appropriate to selected points on the operating range of the process. By this means the effect of variations in process dynamics can be reduced. Although the method can deal with general changes in process dynamics it is known as gain scheduling since the scheme was devised originally to accommodate changes in process gain only. The gain scheduling control system is illustrated in Figure 3.7.

The gain scheduling technique was originally used in aircraft flight control systems and is still widely and successfully applied

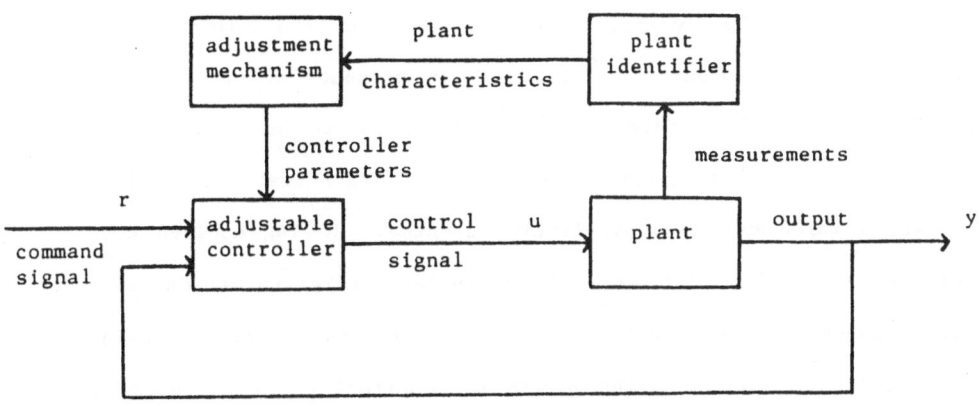

<u>Figure 3.6</u> : Adaptive Feedback
Controller

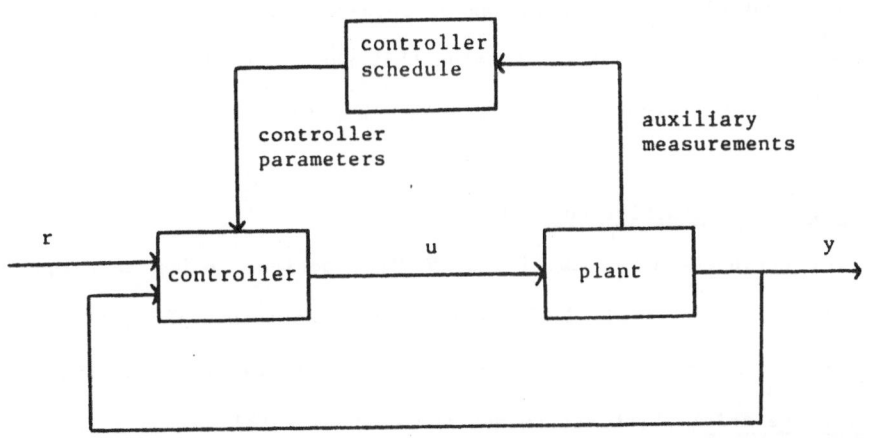

<u>Figure 3.7</u> : Gain Scheduling

in this area: the dynamic characteristics of the aircraft are inferred from measurements of the dynamic pressure and air density (Parks et al 1980). Gain scheduling can also be used in process control problems where the key problem is to find suitable scheduling variables which can be conveniently measured. Having decided upon suitable scheduling variables the controller parameters are obtained for a number of points on the operational range of the process using some design technique. The particular controller setting used at any given time is then selected according to continual measurement of the physical scheduling variable.

A drawback of gain scheduling is that suitable controller parameters for a range of operating conditions must be obtained in advance of process operation. Depending on the number of operating conditions for which the controller must be designed, this can be a time-consuming procedure as the performance and stability features of each design must be satisfactory. An advantage of gain scheduling is that the controller parameters can be changed quickly in response to changes in process dynamics, assuming that the appropriate scheduling variables can be measured accurately and quickly.

Model reference adaptive control

The aerospace problems of the 1950's prompted another technique for automatic adjustment of the controller parameters. The design of model reference adaptive controllers consists of the specification of a reference model which determines the desired ideal response of the process output to the command signal. The model-reference adaptive control method is illustrated in Figure 3.8. The system consists of a normal feedback control loop together with an outer loop which continually adjusts the controller parameters in an attempt to make the process output y(t) the same as the reference-model output $y_m(t)$. The adjustment mechanism is driven by the error $e_m(t)$ between the ideal model output $y_m(t)$ and the actual process output y(t). The adjustment mechanism is designed in such a way that the controller parameters are altered so as to make this error as close as possible to zero.

The original mechanism proposed for the adjustment of controller parameters was developed in the late 1950's by workers at Massachussetts Institute of Technology in the USA in connection with the design of aircraft flight control systems (Whitaker et al, 1958). The adjustment rule, which subsequently became known as the 'MIT rule', is given by the heuristic law

$$\frac{d\theta}{dt} = -ke_m \text{grad}_\theta e_m \qquad (3.10)$$

In this equation e_m is again the error between the model output and the actual process output. The components of the vector θ are the adjustable controller parameters and the components of the vector $\text{grad}_\theta e_m$ are the sensitivity derivatives of the error with respect to the adjustable parameters. k is a design parameter which determines the adaptation rate.

The adjustment rule was originally motivated using the following heuristic argument; assume that the controller parameters θ change much more slowly than the other system variables. To drive the error e_m as close to zero as possible the controller parameters are then changed in the direction of the negative gradient of e_m^2.

The MIT-rule can be rewritten in the form

$$\theta(t) = - k \int e_m(s) \text{grad}_\theta e_m(s) ds \qquad (3.11)$$

It can be seen that the adjustment mechanism consists of three parts: a linear filter for computing sensitivity derivatives, a multiplier and an integrator.

The critical aspect of model-reference adaptive control is the stability of the overall system. Subsequent developments of the MIT rule have been obtained using stability theory (Landau, 1979). Several applications of the theory have been reported, most notably in power system control (Irving, 1979) and in ship steering control problems (Van Amerongen, 1981).

Self-tuning control

Although many successful applications of gain scheduling and model-reference adaptive control have been reported, the total of such designs is far outnumbered by the third major class of adaptive algorithms: self-tuning controllers.

Application of the analytical design techniques discussed in a previous section consists of two mains steps: identification of a model of the process and controller design. The success of these designs depends upon the accuracy with which the process dynamics are known. Initial ignorance of, or subsequent variation in, the process dynamics can result in poor control quality.

Self-tuning control is a discrete-time method which attempts to overcome these problems by automating the overall design procedure and repeating the steps of identification and controller design during each sample interval. The self-tuning controller therefore has the ability to tune itself initially and to re-tune should the process dynamics subsequently change. The self-tuning controller is shown in Figure 3.9. The system consists of a normal feedback loop with process and controller, and an outer loop which continually adjusts the controller parameters. The outer loop is composed of two main parts:

(i)   A recursive parameter estimation routine which uses measurements of the process input and output to continually update a model of the process. The model is normally a simple transfer-function.

(ii)   A controller design algorithm which calculates the controller parameters using the latest estimate of the process model.

This approach to self-tuning control is known as an <u>explicit</u> method since the process model itself is explicitly estimated. In some self-tuning algorithms it is possible to re-parameterise the process model such that it is expressed in terms of the controller parameters. In this type of algorithm, known as an <u>implicit</u> method, the controller design step is eliminated since the controller parameters themselves are estimated by the recursive identification routine.

The explicit self-tuning method is very flexible since there is freedom both in the choice of recursive estimation algorithm and controller design method.

Although the self-tuning control concept was first proposed in the late 1950's (Kalman, 1958), the early hardware limitations meant that the first successful applications did not appear until the early

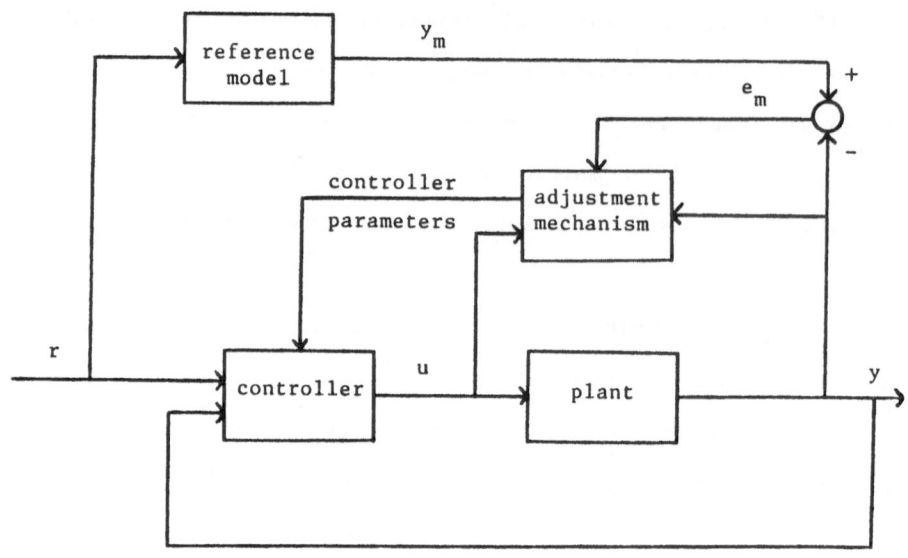

Figure 3.8 : Model-reference
Adaptive Control

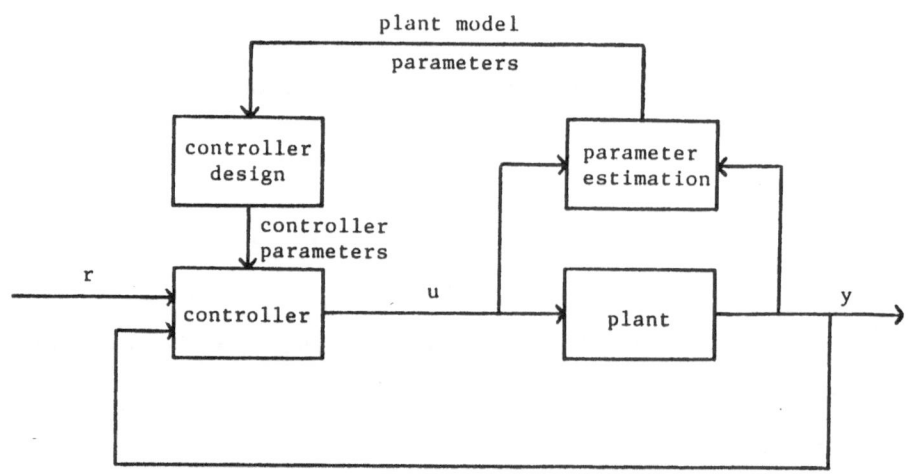

Figure 3.9 : Self-tuning Control

1970's. Since that time the self-tuning control method has proved to be the most successful in applications and commercial products using the technique have appeared. The adaptive algorithm presented in Chapter 4 belongs to the class of self-tuning controllers.

3.7 PREVIOUS APPROACHES TO SELF-TUNING CONTROL

The earliest practical self-tuning control algorithms were based upon stochastic control methods derived using polynomial techniques. Peterka (1970) combined Åström's (1970) minimum-variance control law with a recursive least-squares based parameter estimation algorithm to produce the first version of the celebrated Self-Tuning Regulator (STR). The STR was later studied in depth by Åström and Wittenmark (1973, 1985). A possible source of difficulty in the STR arises when the controlled plant is inverse unstable ('non-minimum phase'). Since the poles of the minimum-variance controller cancel the plant zeros the resulting closed-loop system is unstable whenever the plant is inverse unstable, although this problem may be alleviated in certain cases by careful selection of the sampling period (Åström and Wittenmark, 1985). Peterka (1972) subsequently derived the stable minimum-variance control law for inverse unstable plant and this was studied in the self-tuning control context by Åström and Wittenmark (1974). The stability properties of the minimum-variance control law are a direct result of the fact that no penalty is placed on the magnitude of the control signal generated by the controller.

The Generalised Minimum-Variance (GMV) control law (Clarke and Gawthrop 1975, 1979) evolves from minimisation of a single-stage cost-function which includes control costing. The GMV method was

introduced in an attempt to overcome the stability problems
encountered with the minimum-variance regulator. For inverse
unstable plant, however, closed-loop stability is <u>conditional</u> upon
proper selection of the cost-function weights. The choice of
cost-function weights is particularly difficult when the controlled
plant is also unstable.

The GMV self-tuner has nevertheless been applied in several
industrial problems (Hodgson 1982, Tuffs 1984). Gawthrop (1977) had
earlier given the method added flexibility by showing that,
depending on the choice of cost weights, the method could be
interpreted as providing model-following, detuned model-following or
Smith predictive control.

The Weighted Minimum-Variance (WMV) control law derived by
Grimble (1981) extends the GMV method to plants which may be both
open-loop unstable and inverse unstable. Grimble shows that a stable
closed-loop system may always be achieved if the cost-function
weights are suitably chosen.

The use of Kučera's (1979) polynomial equation approach to LQG
control was first applied in the self-tuning control context by
Zhao-Ying and Åström (1981) and Åström (1983b). Grimble (1984)
derived an <u>implicit</u> LQG self-tuning algorithm. The <u>explicit</u> method
was extended to include dynamic cost-function weights and feedforward
control of measurable disturbances by Hunt et al (1986, 1987) and
Hunt and Grimble (1988) and is fully treated in the following
chapter. The LQG controllers have a guarantee of closed-loop
stability regardless of the plant pole/zero locations.

Self-tuning LQG controllers based upon the standard state-space formulation have been considered by Lam (1980) and Clarke et al (1985). Peterka (1986) has derived an LQG self-tuner which is applicable to both the ARMA and difference operator (Goodwin, 1985) model forms. In Peterka's approach algorithmic and numerical aspects are emphasised and the final controller design is obtained using state-space transformations. Peterka's method has been permanently installed in several industrial applications (see, for example, Ettler 1986 and Lizr 1986).

A further class of self-tuning control methods which has received growing attention in recent years is the family of long-range predictive controllers. The first attempts to use long-range prediction concepts in controller design were proposed in the IDCOM method of Richalet et al (1978) and the DMC algorithm of Cutler and Ramaker (1980). A unifying idea in the long-range predictive methods is to extend the prediction horizon beyond non-minimum phase effects and time delays. Self-tuning controllers based on this idea have been proposed by Ydstie (1984), Peterka (1984), Mosca et al (1984) and De Keyser and Van Cauwenberghe (1985). The Generalised Predictive Control (GPC) method of Clarke et al (1987) effectively extends the GMV method by use of long-range prediction over a multi-stage cost-function, and can overcome the problem of stabilising non-minimum phase plant if the prediction horizon is chosen to be long enough. The GPC method can also be given the same design polynomials and interpretations discussed by Gawthrop (1977) for the GMV algorithm.

One final class of self-tuning controllers is the continuous-time approach studied by Egardt (1979b) and extensively developed by

Gawthrop (1986). The method proposed by Gawthrop is based upon the premise that robust adaptive controllers will arise by basing the design upon established control engineering principles and practice. In this spirit, the design is performed in the continuous-time domain although the implementation is still digital. Egardt and Gawthrop also demonstrate that a number of different algorithms can be unified in the continuous-time framework.

3.8 CONTRIBUTIONS OF THE PRESENT WORK

As mentioned above, there are several approaches to the design of self-tuning control systems using stochastic optimal control theory ('LQG' control). The method followed in Chapter 4 uses the theory developed in Chapter 2 which is based upon Kučera's (1979) polynomial equation approach. In this approach the design procedure reduces to the solution of polynomial equations whose coefficients are obtained by spectral factorisation. These equations can be solved using fast and efficient numerical algorithms.

The polynomial approach to LQG control offers a flexible design method which can be readily used as the basis of a self-tuning control algorithm. The new algorithm presented in Chapter 4 contains some refinements and important extensions of the earlier work by Grimble (1984) and Hunt et al (1986). In particular, the optimal tracking problem in the presence of a measurable disturbance was solved in Chapter 2. The solution naturally involves the use of a feedforward compensator and is described in Chapter 4 in the self-tuning control framework.

Use of the polynomial equation solution to the optimal

feedforward control problem in a self-tuning control algorithm was first given by Hunt, Grimble and Jones (1986, 1987) and is also studied in Hunt and Grimble (1988). Sternad (1987) has made a very detailed and independent study of the optimal feedforward technique applied to both fixed-parameter and self-tuning systems. Sternad also compares the optimal technique using polynomial methods with previous approaches to the feedforward compensation of measurable disturbances.

A further feature of the design method presented is the use of frequency-dependent weighting elements in the cost-function. The dynamic weights allow the frequency-response of the closed-loop system to be shaped in a straightforward manner.

The optimal control law which is presented in Chapter 4 is a simplified version of the theory derived in Chapter 2. In particular, all measurement noises are assumed to be zero in order to reduce the plant model to the basic ARMAX form which is sufficient for general-purpose self-tuning algorithms (the full complexity of the model used in Chapter 2 is required in some specialised applications such as ship steering (Grimble, 1986a) where, for identification purposes, a priori knowledge about the coloured measurement noise sub-system is available in the form of standard wave spectra).

The two-degrees-of-freedom (2DF) plus feedforward optimal controller which was derived in Section 2.7 is employed for the LQG self-tuner. The optimal controller consists of three parts (feedback, reference and feedforward) which process the system output, reference and measurable disturbance signals separately.

Three possible design strategies are proposed:

(i) The complete general solution of the optimal control problem which involves three couples of polynomial equations, one couple being associated with each part of the controller.

(ii) Each couple of polynomial equations can be reduced to a single 'implied' equation. Under certain stated conditions the three implied polynomial equations can be solved to obtain the unique optimal controller. Solution of the implied equations is computationally simpler than solution of the original couples.

(iii) In the optimal control design the feedback part of the controller is independent of the reference and feedforward parts. A third design strategy is proposed in which the optimal feedback controller is calculated in the normal way and then the reference and feedforward parts are calculated non-optimally using steady-state considerations. The method is useful in situations where the available computation time is short as the reference and feedforward polynomial equations no longer need to be solved.

The robustness properties of the LQG self-tuner are discussed by summarising the important features of the control design and various techniques which have been used to achieve robust parameter estimation. The main results of a recent convergence analysis (Grimble, 1986c) are also presented in Chapter 4.

Chapter 4 concludes with a discussion of practical issues relating to control law implementation, cost-function weight selection and computational issues. Proofs of all the results stated in Chapter 4 can be found in Chapter 2 (with the exception of the convergence proof which is due to Grimble, 1986c).

CHAPTER FOUR

OPTIMAL SELF-TUNING ALGORITHM

Summary

The open-loop model for the single-input single-output plant
under consideration is described in Section 4.1. The plant output
which is to be controlled is affected by two disturbance signals, one
of which is assumed measurable.

In Section 4.2 the two-degrees-of freedom (2DF) controller
structure employed is introduced. In addition, a feedforward
compensator is used to reject the measurable disturbance signal.
The optimal controller consists of three parts (feedback, reference
and feedforward) which process the system output, reference and
measurable disturbance signals separately. Three possible design
strategies are proposed in Section 4.2:

(i)   The complete general solution of the optimal control
        problem (Section 4.2.1).

(ii)   The optimal solution using the 'implied' polynomial
        equations. The conditions under which the implied
        equations yield the unique optimal controller are stated
        (Section 4.2.2).

(ii)   A computationally simpler design where the feedback part
        is calculated optimally, and the reference and
        feedforward parts of the controller are calculated to
        give correct steady state performance (Section 4.2.3).

The robustness properties of the LQG self-tuner are discussed in
Section 4.4 by summarising the important features of the control

design and various techniques which have been used to achieve robust parameter estimation.   The main results of a recent convergence analysis are presented in Section 4.5.

The chapter concludes in Section 4.6 with a discussion of practical issues relating to control law implementation, cost-function weight selection and computational issues.

156

4.1 MODEL STRUCTURE

The open-loop model for the single-input single-output <u>plant</u> under consideration is shown in Figure 4.1. The plant is governed by the equation:

$$y(t) = p(t) + x(t) + d(t) \tag{4.1}$$

$$= W_p u(t) + W_x \ell(t) + W_d \psi_d(t) \tag{4.2}$$

where $y(t)$ is the output to be controlled, $u(t)$ is the plant control input, $\psi_d(t)$ is an unmeasurable disturbance, and $\ell(t)$ is a disturbance which is available for measurement. Denoting the least-common-denominator of W_p, W_x and W_d as A, these sub-systems may be expressed as:

$$W_p = A^{-1}B \tag{4.3}$$

$$W_d = A^{-1}C \tag{4.4}$$

$$W_x = A^{-1}D \tag{4.5}$$

where A,B,C and D are polynomials in the delay operator d.

Reference generator

The system output $y(t)$ is required to follow as closely as possible a reference signal $r(t)$. The signal $r(t)$ is represented as the output of a generating sub-system W_r which is driven by an external stochastic signal $\psi_r(t)$:

$$r(t) = W_r \psi_r(t) \tag{4.6}$$

The sub-system W_r is represented in polynomial form as:

$$W_r = A_e^{-1} E_r \tag{4.7}$$

where A_e and E_r are polynomials in d.

The tracking error e(t) is defined as:

$$e(t) \underset{=}{\Delta} r(t) - y(t) \qquad\qquad (4.8)$$

Any common factors of A_e and A are denoted by D_e such that:

$$A_e = D_e A'_{ec} \ , \ A = D_e A' \qquad\qquad (4.9)$$

Measurable disturbance generator

The measurable disturbance signal $\ell(t)$ may be represented as the output of a generating sub-system W_ℓ driven by an external stochastic signal $\psi_\ell(t)$;

$$\ell(t) = W_\ell \psi_\ell(t) \qquad\qquad (4.10)$$

The sub-system W_ℓ is represented in polynomial form as:

$$W_\ell = A_\ell^{-1} E_\ell \qquad\qquad (4.11)$$

where A_ℓ and E_ℓ are polynomials in d.

Assumptions

1. Each sub-system is free of unstable hidden modes.

2. The plant input-output transfer-function W_p is assumed strictly causal i.e. <B> = 0.

3. The polynomials A and B must have no unstable common factors.

4. Any unstable factors of A_e must also be factors of A.

5. Any unstable factors of A_ℓ must also be factors of both A and D.

6. The polynomials C, E_r and E_ℓ may, without loss of generality, be assumed stable.

These assumptions, together with the assumptions on the cost-function weighting elements given in the following section, amount to the solvability conditions for the optimal control problem (see Theorem 14). By making these assumptions, therefore, we ensure that a solution to any given problem exists. The question of artificially ensuring problem solvability is discussed in Section 4.6.3.

Different types of reference and disturbance signals ($r(t)$, $\ell(t)$ and $d(t)$ in Figure 4.1) may be admitted by appropriate definition of the external stochastic signals, $\psi_r(t)$, $\psi_\ell(t)$ and $\psi_d(t)$, namely:

(i)   Coloured zero-mean signals are generated when the driving source (ψ_r, ψ_ℓ or ψ_d) is a zero-mean white noise sequence and the filter (W_r, W_ℓ or W_d) is asymptotically stable.

(ii)   Random walk sequences are generated when the driving source is a zero-mean white noise sequence and the filter has a denominator 1-d.

(iii) Step-like sequences consisting of random steps at random times are generated when the driving source is a Poisson process and the filter has a denominator 1-d.

(iv)   Deterministic sequences (such as steps, ramps or sinusoids) are generated when the driving source is a unit pulse sequence and the filter has poles on the unit circle of the d-plane.

4.2 CONTROLLER DESIGN

In the closed-loop system a two-degrees-of-freedom (2DF) control structure is used. In addition, a feedforward compensator is employed to counter the effect of the measurable disturbance $\ell(t)$.

Controller structure

The closed-loop system is shown in Figure 4.2. The control law is given by:

$$u(t) = -C_{fb}y(t) + C_r r(t) - C_{ff}\ell(t) \qquad (4.12)$$

where the feedback controller C_{fb}, the reference controller C_r, and the feedforward controller C_{ff} may be expressed as ratios of polynomials in the delay operator d as:

$$C_{fb} = C_{fbd}^{-1}C_{fbn} \qquad (4.13)$$

$$C_r = C_{rd}^{-1}C_{rn} \qquad (4.14)$$

$$C_{ff} = C_{ffd}^{-1}C_{ffn} \qquad (4.15)$$

Cost function

The desired optimal controller evolves from minimisation of the cost-function:

$$J = E\{(H_q e)^2(t) + (H_r u)^2(t)\} \qquad (4.16)$$

where H_q and H_r are dynamic (i.e. frequency dependent) weighting elements which may be realised by rational transfer-functions.

Using Parseval's Theorem the cost-function may be transformed to the frequency domain and expressed as:

$$J = \frac{1}{2\pi j} \oint_{|z|=1} \{Q_c\phi_e + R_c\phi_u\} \frac{dz}{z} \qquad (4.17)$$

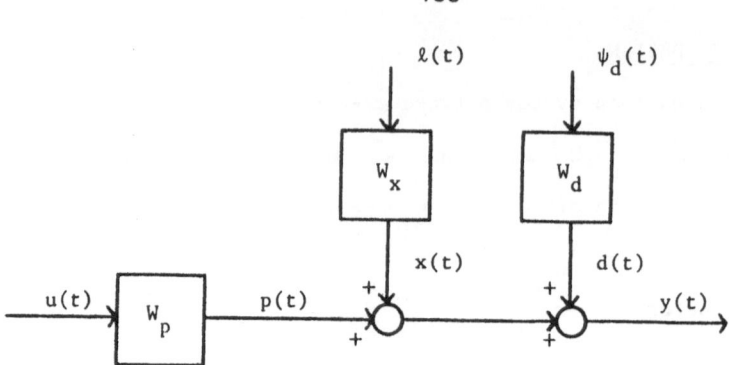

Figure 4.1 : Plant model

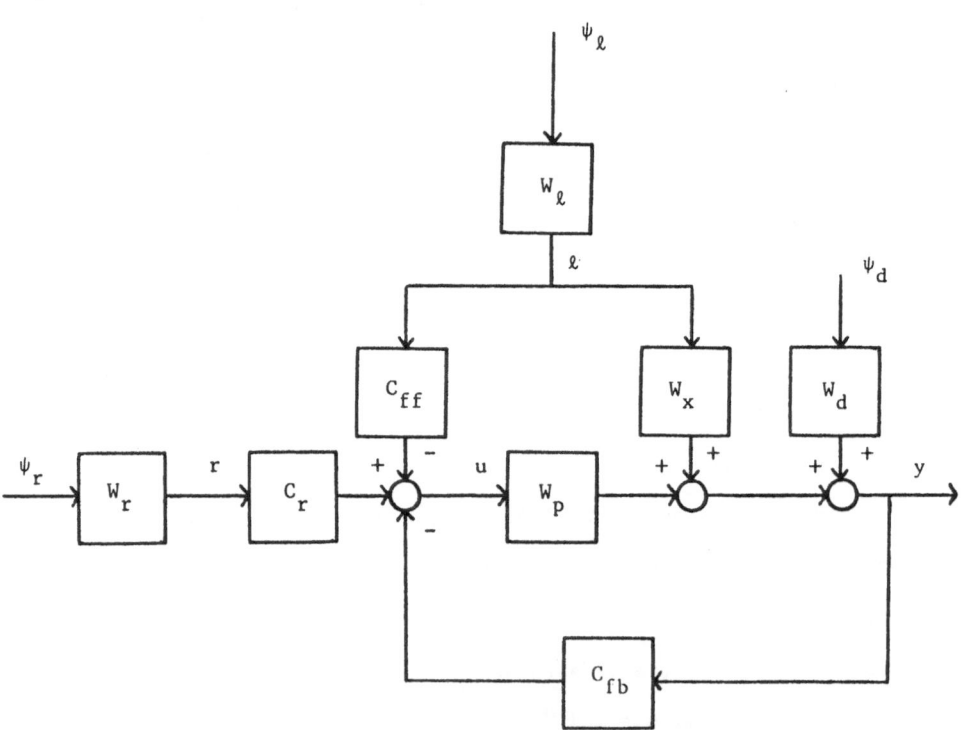

Figure 4.2 : Closed-loop System

where ϕ_e and ϕ_u are the tracking error and control input spectral densities, respectively, and:

$$Q_c = H_q H_q^* \; , \; R_c = H_r H_r^* \tag{4.18}$$

The weighting elements Q_c and R_c may be expressed as ratios of polynomials in the delay operator d as:

$$Q_c \triangleq \frac{B_q^* B_q}{A_q^* A_q} \; , \; R_c \triangleq \frac{B_r^* B_r}{A_r^* A_r} \cdot \tag{4.19}$$

Assumptions

1. The weighting elements Q_c and R_c are <u>strictly positive</u> on $|d|=1$.

2. A_q, B_q, B_r and A_r are <u>stable</u> polynomials.

4.2.1 Optimal Control Law

The <u>stable</u> spectral factor D_c is defined by:

$$D_c^* D_c = B^* A_r^* B_q^* B_q A_r B + A^* A_q^* B_r^* B_r A_q A \tag{4.20}$$

The feedback, reference and feedforward parts of the control law (4.12) which minimises the cost-function (4.17) are as follows:

(i) Optimal feedback controller

$$C_{fb} = \frac{GA_r}{HA_q} \tag{4.21}$$

where G, H (along with F) is the solution having the property:

$$(D_c^* z^{-g})^{-1} F \text{ strictly proper}$$

of the polynomial equations:

$$D_c^* z^{-g} G + FAA_q = B^* A_r^* B_q^* B_q Cz^{-g} \tag{4.22}$$

$$D_c^* z^{-g} H - FBA_r = A^* A_q^* B_r^* B_r Cz^{-g} \tag{4.23}$$

where $g > 0$ is the smallest integer which makes the equations (4.22)-(4.23) polynomial in d.

(ii) <u>Optimal reference controller</u>

$$C_r = \frac{MA_r C}{E_r HA_q} \qquad (4.24)$$

where M (along with N and Q) is the solution having the property:

$$(D_c^* z^{-g})^{-1} N \quad \text{strictly proper}$$

of the polynomial equations:

$$D_c^* z^{-g} M + NA_q A_e = B^* A_r^* B_q^* B E_r z^{-g} \qquad (4.25)$$

$$D_c^* z^{-g} Q - NBA_r A_{ec}' = A^* A_q^* B_r^* B A' E_r z^{-g} \qquad (4.26)$$

(iii) <u>Optimal feedforward controller</u>

$$C_{ff} = \frac{A_r (XC - GE_\ell D)}{HA_q E_\ell A} \qquad (4.27)$$

where X (along with Z and Y) is the solution having the property:

$$(D_c^* z^{-g})^{-1} Z \quad \text{strictly proper}$$

of the polynomial equations:

$$D_c^* z^{-g} X + ZAA_q A_\ell = B^* A_r^* B_q^* B DE_\ell z^{-g} \qquad (4.28)$$

$$D_c^* z^{-g} Y - ZBA_r A_\ell = A^* A_q^* B_r^* B DE_\ell z^{-g} \qquad (4.29)$$

4.2.2 Implied Diophantine Equations

The general solution of the optimal control problem involves three couples of polynomial equations, one couple for each part of the control law. However, under certain conditions each couple can be replaced by a single, related, equation. Optimality of the **implied** diophantine equations requires the following additional assumptions:

Assumptions

1. The disturbance sub-systems $A^{-1}C$, $A^{-1}D$, $A_e^{-1}E_r$ and $A_\ell^{-1}E_\ell$ are assumed to be proper rational transfer-functions.

2. The cost-function terms $A_q^{-1}B_q$ and $A_r^{-1}B_r$ are assumed to be proper rational transfer-functions.

3. The polynomial pairs A_q,A_r, A_q,B and A_r,A are each assumed to be coprime.

We can easily ensure that Assumptions 2 and 3 are satisfied by appropriate choice of the cost-function weights.

(i) Implied feedback equation

The polynomials G and H in equations (4.22)-(4.23) also satisfy the polynomial equation:

$$AA_q H + BA_r G = D_c C \qquad (4.30)$$

The optimal feedback controller polynomials G and H are determined uniquely by the solution of equation (4.30) having the property:

$$(AA_q)^{-1}G \quad \text{strictly proper}$$

iff the polynomials A and B are coprime.

(ii) Implied reference equation

The polynomials M and Q in equations (4.25)-(4.26) also satisfy the polynomial equation:

$$D_e A_q Q + BA_r M = D_c E_r \qquad (4.31)$$

The optimal reference controller polynomial M is determined

uniquely by the solution of equation (4.31) having the property:

$$(A_e A_q)^{-1} M \quad \text{strictly proper}$$

iff the polynomials A and B are coprime <u>and</u> A_e is a divisor of A . When this condition holds then, from equation (4.9), $D_e = A_e$ and the implied equation (4.31) becomes:

$$A_e A_q Q + BA_r M = D_c E_r \qquad (4.32)$$

(iii) Implied feedforward equation

The polynomials X and Y in equations (4.28)-(4.29) also satisfy the polynomial equation:

$$AA_q Y + BA_r X = D_c DE_\ell \qquad (4.33)$$

Assume now that A_ℓ is a divisor of both A and D. From equations (4.28)-(4.29) A_ℓ must then also divide both X and Y. The implied feedforward diophantine equation (4.33) becomes:

$$AA_q Y' + BA_r X' = D_c D' E_\ell \qquad (4.34)$$

where:

$$Y \triangleq A_\ell Y', \quad X \triangleq A_\ell X', \quad D \triangleq A_\ell D' \qquad (4.35)$$

The conditions for optimality of the implied feedforward diophantine equation may now be stated as follows: the optimal feedforward controller polynomial X is given by $X = A_\ell X'$ where X' is determined <u>uniquely</u> by the solution of equation (4.34) having the property:

$$(AA_q)^{-1} X' \quad \text{strictly proper}$$

iff the polynomials A and B are coprime <u>and</u> the polynomial A_ℓ is a divisor of both A and D.

Discussion

The condition that the A and B polynomials must be coprime is required in order that the implied feedback, reference and feedforward diophantine equations uniquely determine the optimal feedback, reference and feedforward parts of the controller. This condition means that all the poles of the disturbance sub-systems W_d and W_x must also be poles of the plant input-output transfer-function W_p.

The extra condition required for optimality of the implied reference equation is that A_e, the reference generator denominator, must divide A. In the case of the unstable reference generators of greatest practical interest (such as steps, ramps etc) this condition corresponds to one of the optimal control problem solvability conditions (see Assumption 4 in Section 4.1). If, therefore, the reference generator is unstable and the optimal control problem is solvable (by satisfying assumptions 1-6 in Section 4.1 and Assumptions 1-2 in Section 4.2 we can ensure that the problem is solvable) then the condition for optimality of the implied reference diophantine equation reduces to the condition that A and B must be coprime.

The conditions required for optimality of the feedforward diophantine equation are that A and B must be coprime and that A_ℓ, the measurable disturbance generator denominator, must divide both A and D. For the unstable disturbance generators of practical importance the condition that A_ℓ must divide A and D again corresponds to one of the optimal control problem solvability conditions (see Assumption 5 in Section 4.1). If, therefore, the measurable disturbance generator is unstable and the optimal control problem is solvable then the condition for optimality of the implied feedforward diophantine equation reduces to the condition that A and B must be coprime (this is the same condition required for optimality of the implied feedback diophantine equation).

The question of ensuring problem solvability is discussed in Section 4.6.3.

4.2.3 Simplified Design

The control structure used allows flexibility in the design of the reference and feedforward parts of the controller. Having designed the optimal feedback controller the reference and feedforward parts can then, if desired, be designed independently of the feedback properties of the system. This option may be important in cases where the full LQG design cannot be implemented due to computational constraints. In such cases the reference and feedforward parts of the controller can be designed to ensure proper steady-state performance as follows:

(i) Reference controller

The reference controller may be defined as:

$$C_r = \frac{\gamma A_r C}{H A_q} \qquad (4.36)$$

where the _scalar_ γ replaces the polynomial M in equation (4.24). From the closed-loop system structure shown in Figure 4.2 the transfer-function between the reference signal $r(t)$ and the controlled output $y(t)$ may be calculated as:

$$T_{y/r} = \frac{\gamma B}{D_c} \qquad (4.37)$$

The scalar γ is chosen so as to ensure unity steady-state gain between $r(t)$ and $y(t)$ as:

$$\gamma = \frac{D_c(1)}{B(1)} \qquad (4.38)$$

Shaping of the command response may also be achieved by cascading a unity-gain shaping filter with the reference controller.

(ii) Feedforward controller

From the closed-loop system structure shown in Figure 4.2 the transfer-function between the measurable disturbance $\ell(t)$ and the controlled output $y(t)$ may be calculated as:

$$T_{y/\ell} = \frac{C_{fbd}(DC_{ffd} - BC_{ffn})}{C_{ffd}(AC_{fbd} + BC_{fbn})} \qquad (4.39)$$

The effect of $\ell(t)$ on the output is eliminated when:

$$DC_{ffd} - BC_{ffn} = 0 \qquad (4.40)$$

or, when the feedforward controller is defined as:

$$C_{ff} = \frac{D}{B} \qquad (4.41)$$

There are, however, two major problems with this design:

(i) Whenever the delay associated with the measurable disturbance filter D/A is less than the delay in the plant input-output transfer-function B/A a non-causal controller results.

(ii) The feedforward controller is unstable whenever the plant input-output transfer-function B/A is inverse unstable i.e. when B(d) has zeros inside the unit circle of the d-plane.

A solution to these problems is to sacrifice the transient

properties of the feedforward controller in favour of a static
feedforward design which is calculated to ensure the elimination of
the measurable disturbance in steady-state:

$$C_{ff} = \frac{D(1)}{B(1)} \qquad (4.42)$$

The two problems of the non-optimal design outlined above demonstrate
a clear advantage of the optimal feedforward design since a causal
and stable feedforward controller will always result in the optimal
design regardless of the relative delays in the B and D polynomials,
and regardless of the positions of the zeros of B.

4.3 LQG SELF-TUNING CONTROL ALGORITHM

An explicit LQG self-tuning controller may be constructed using the certainty equivalence argument, where the A,B,C and D polynomials in the LQG design of Section 4.2 are replaced by their estimated values $\hat{A},\hat{B},\hat{C}$ and $\hat{D}$.

The explicit LQG self-tuning control algorithm may be summarised as follows:

Step 1 : Choose cost-function weights.

Step 2 : Update estimates of A,B,C and D polynomials.

Step 3 : Solve spectral factorisation (4.20) for D_c.

Step 4 : Solve equations (4.22)-(4.23) for G and H, and form feedback controller according to equation (4.21).

Step 5 : Solve equations (4.25)-(4.26) for M, and form reference controller according to equation (4.24).

Step 6 : Solve equations (4.28)-(4.29) for X, and form feedforward controller according to equation (4.27).

Step 7 : Calculate and implement new control signal.

Step 8 : Goto Step 1 at next sample instant.

In steps 4-6 it may be possible to solve the implied diophantine equations to obtain the optimal controller polynomials, as discussed in Section 4.2.2. Similarly, in steps 5-6 it may be desirable to use the steady-state designs outlined in Section 4.2.3.

4.4 ROBUSTNESS OF THE LQG SELF-TUNER

The subject of robustness of self-tuning controllers is one which has generated a great deal of discussion and controversy in recent years. This discussion has largely been stimulated by the paper of Rohrs et al (1982) which analysed the robustness properties of the self-tuning regulator (STR) and of model reference adaptive controllers (MRAC) in the presence of unmodelled dynamics and disturbances. These authors concluded that adaptive controllers are inherently non-robust and this stimulated a very active debate leading to some useful insights into the robustness question.

Åström (1983c) directly challenged the allegations of Rohrs, while Goodwin et al (1985) pointed out that the approach of analysing existing high performance adaptive controllers would almost certainly reveal poor robustness properties.

Åström (1983b) and Goodwin et al (1985) take a more pragmatic approach to the robustness issue by attempting to obtain a robust adaptive controller by combining a robust control law with a robust identification algorithm.

In the following discussion the robustness properties of the LQG design are examined and methods of achieving robust parameter estimation are discussed.

4.4.1 Robustness of the LQG Design

The performance and properties of feedback control systems have long been understood by control engineers in terms of the frequency responses of the various system transfer-functions (Truxal 1955, Horowitz, 1963). The main ideas of conventional frequency-response design methods have recently been supported by theoretical analysis

(Doyle and Stein, 1981).

These ideas can be summarised with reference to a typical Bode plot of the loop gain as shown in Figure 4.3. It is well understood that there are three important frequency regions:

(i)   The low-frequency region where the loop gain should be high to achieve good command response, disturbance rejection and robust performance properties.

(ii) The crossover region where the stability margins should be adequate.

(iii) The high-frequency region where the loop gain should fall off rapidly to achieve robust stability (i.e. insensitivity to unmodelled high-frequency dynamics) and insensitivity to measurement noise.

Any competent design of a digital control system should include anti-aliasing filters to eliminate signal transmission above the Nyquist frequency (Åström and Wittenmark, 1984). The high-frequency properties of the controller will therefore depend critically upon the sampling period.

The relevant features of the LQG controller may be investigated by summarising the properties of the design presented in Section 4.2:

(i)   The feedback, reference and feedforward parts of the controller each have poles due to the A_q weighting term, and zeros due to the A_r weighting term (see equations (4.21), (4.24) and (4.27)). Thus, loop-shaping may easily be achieved by manipulation of the cost function weights. In particular, the desired high gain at low-frequency can be achieved by introducing integral action when the A_q term is chosen as $A_q = 1-\nabla d$, $\nabla \rightarrow 1$.

A particularly simple formulation of the weighting terms Q_c and R_c which requires the selection of only two parameters is presented in Section 4.6.2.

(ii) The stability margins for the closed-loop system can be examined using the implied feedback diophantine equation (4.30):

$$AA_q H + BA_r G = D_c C \triangleq T \qquad (4.43)$$

From the feedback controller equation (4.21) and the closed-loop system model in Figure 4.2 it may easily be verified that equation (4.43) is the characteristic equation of the closed-loop system, where T is defined as the characteristic polynomial. This shows that the nominal closed-loop system is guaranteed to be stable, since the polynomials D_c and C are by definition stable.

This result should be contrasted with the stability properties of the Self-Tuning Regulator (Åström and Wittenmark, 1973) and the Self-Tuning Controller (Clarke and Gawthrop, 1975, 1979). It is possible that the closed-loop system for these control laws will be nominally unstable, particularly when the controlled process has zeros inside the unit circle in the d-plane.

Equation (4.43) also shows that standard pole-assignment may be obtained as a by-product of the LQG algorithm by solving the equation:

$$AA_q H + BA_r G = T_c C \qquad (4.44)$$

where T_c is chosen as the desired closed-loop pole polynomial.

Notice that in this formulation of the pole-assignment algorithm the loop-shaping properties of the LQG design are retained since the A_q and A_r polynomials remain in equation (4.44). Use of the pole-assignment algorithm introduces computational savings since the spectral factorisation (4.20) is no longer required, and in subsequent calculations the polynomial D_c is replaced by T_c.

4.4.2 Robust Parameter Estimation

In this section the recursive Extended-Least-Squares (ELS) estimation algorithm is described and the methods which can be used to achieve robust parameter tracking are briefly reviewed. The plant model (4.2) is re-written in the approximate form:

$$y(t) = \underline{\phi}^T(t)\underline{\theta}(t) + \psi_d(t) \tag{4.45}$$

where the <u>parameter vector</u> $\underline{\theta}(t)$ and <u>regression vector</u> $\underline{\phi}(t)$ are defined by:

$$\underline{\theta}^T(t) = \left[a_1 \cdots a_{na}; \ b_o \cdots b_{nb}; \ d_o \cdots d_{nd}; c_1 \cdots c_{nc}\right] \tag{4.46}$$

$$\underline{\phi}^T(t) = \left[-y(t-1) \cdots -y(t-na); u(t-k_1) \cdots u(t-k_1-nb);\right.$$
$$\left.\ell(t-k_2) \cdots \ell(t-k_2-nd); v(t-1) \cdots v(t-nc)\right] \tag{4.47}$$

The parameters, $a.,b.,c.$ and $d.$ are the coefficients of the polynomials A, B, C and D. k_1 and k_2 represent the time-delays in the sub-systems W_p and W_x as integer multiples of the sample period. The signal $v(t)$ is a proxy to the unmeasurable signal $\psi_d(t)$, defined by:

$$v(t) = y(t) - \underline{\phi}^T(t)\underline{\hat{\theta}}(t) \tag{4.48}$$

where $\underline{\hat{\theta}}(t)$ is the vector of <u>estimated</u> parameters.

The recursive ELS algorithm may be summarised as follows:

$$P(t) = \frac{1}{\lambda(t)} \left(P(t-1) - \frac{P(t-1)\underline{\phi}(t)\underline{\phi}^T(t)P(t-1)}{\lambda(t) + \underline{\phi}^T(t)P(t-1)\underline{\phi}(t)}\right) \tag{4.49}$$

$$\underline{k}(t) = P(t)\underline{\phi}(t) = \frac{P(t-1)\underline{\phi}(t)}{\lambda(t) + \underline{\phi}^T(t)P(t-1)\underline{\phi}(t)} \qquad (4.50)$$

$$v(t) = y(t) - \underline{\phi}^T(t)\underline{\hat{\theta}}(t) \qquad (4.51)$$

$$\underline{\hat{\theta}}(t) = \underline{\hat{\theta}}(t-1) + \underline{k}(t)v(t) \qquad (4.52)$$

In the above algorithm the 'forgetting factor' $\lambda(t)$ (where $0 < \lambda(t) < 1$) weights the measurements, such that a measurement received n samples ago will have a weighting proportional to λ^n (assuming a constant forgetting factor $\lambda(t) = \lambda$).

The constant forgetting factor technique for parameter tracking has frequently been used in self-tuning control algorithms. The method, however, has some potential implementation difficulties:

(i)   If the algorithm is to remain capable of tracking sudden parameter changes the updating gain $\underline{k}$ must be prevented from becoming too small as the parameter estimates improve. Moreover, if good data is arriving and $\underline{k}$ becomes small, equation (4.49) implies that P is near singular and roundoff error over many updating steps may cause the computed P to become indefinite and the algorithm to break down (the UD factorisation technique (Bierman, 1977) is normally used to alleviate this problem). Thus, λ must not stay too close to unity.

(ii) On the other hand, when λ is less than 1 and little new information on $\underline{\theta}$ is being brought in by the observations, equation (4.49) shows that P may increase as λ^{-1} (the well known 'burst phenomenon', sometimes also known as 'estimator wind-up'). If P becomes large in this way then observation noise, or a sudden increase in information, may induce large spurious variations in $\underline{\theta}$.

When using the constant forgetting factor method, therefore, choice of λ is a difficult and often unsatisfactory compromise. Many methods of adjusting λ(t) automatically in the recursion have been devised (Åström 1980, Fortescue et al 1981, Wellstead and Sanoff, 1981). Alternatively, P may be adjusted directly. For instance, a constant matrix, which can be interpreted as the covariance of random increments in the parameters, may be added at each step then some upper bound applied to P, or the new P may be formed as a weighted sum of the old P and the identity matrix I, the weights being chosen to keep trace P constant (Goodwin et al, 1985). Other methods are described by Egardt (1979), Hägglund (1983), Kulhavy and Karny (1984) and Andersson (1985).

Chen and Norton (1987) have recently described a new parameter tracking method which enables the recursive ELS algorithm to track both abrupt and smooth parameter changes. The method differs from methods based on a scalar forgetting factor by its use of vectors to detect parameter variation, which then results in the relevant element in the updating gain vector being incremented. It also incorporates a test to determine when parameter updating should be suspended so as to avoid divergence when little new information about the parameters is arriving.

The algorithm does not, therefore, have the drawbacks associated with the constant forgetting factor method. The algorithm also embodies one of the key ideas behind robust estimation, namely that estimation should only be performed when 'good' data is arriving.

The method has been described in the context of LQG self-tuning control by Hunt et al (1986).

4.5   CONVERGENCE PROPERTIES

One of the key theoretical problems in self-tuning control which
has received growing attention in recent years is convergence
analysis.   Chen and Caines (1984) and Moore (1984) have tackled the
problem for state-equation based LQG algorithms.

A global convergence result for explicit polynomial based LQG
self-tuning control algorithms of the type under discussion in this
chapter has recently been derived and will be summarised in the
following.   The result relates to the regulator case (i.e.
$r(t) = \ell(t) = 0$) and to a stochastic approximation type
identification algorithm.   The result is due to Grimble (1986c).

To guarantee the convergence properties of the algorithm the
following assumptions must be made:

Assumptions

1.   Upper bounds $n_a = n$, $n_b = m$ and $n_c = q$ on the polynomials A,B
and C are known.

2.   The polynomial $C - k'/2$ is input strictly passive (strictly
positive real) for some real positive constant k'.

3.   There exists a finite T_2 such that $\hat{C}(t;d)$ remains stable for all
$t < T_2$.

4.   Any common roots of $\hat{A}(t;d)$ and $\hat{B}(t;d)$ are strictly and uniformly
outside the unit circle of the d-plane as $t \to \infty$, with
probability one.

Remarks

(a)   Assumption 1 is generally valid and has the useful
property that the transport delay need not be known
exactly.

(b)   A positive real condition such as Assumption 2 often

         arises as a sufficient condition for the convergence of

         recursive parameter estimation schemes.

The global convergence theorem for the explicit LQG self-tuner

may now be stated:

Theorem

Subject to Assumptions 1-4 the explicit LQG self-tuning

regulator, using a stochastic approximation identification algorithm,

is globally convergent in the following sense:

(i) $\displaystyle \lim_{T\to\infty} \sup \frac{1}{T} \sum_{t=1}^{T} y^2(t) < \infty$ a.s.   (4.53)

(ii) $\displaystyle \lim_{T\to\infty} \sup \frac{1}{T} \sum_{t=1}^{T} u^2(t) < \infty$ a.s.   (4.54)

(iii) $\displaystyle \lim_{T\to\infty} \frac{1}{T} \sum_{t=1}^{T} E\left[(y(t)-\hat{y}(t/t-1))^2/F_{t-1} \right] = 1$ a.s.   (4.55)

(iv)   The closed-loop characteristic polynomial $T(t;d)$ converges

       in the sense:

$$\lim_{T\to\infty} \frac{1}{T} \sum_{t=1}^{T} E\left[T(t;d)y(t)-H(t;d)\hat{C}(t-1;d)\varepsilon(t) \right]^2 = 0 \quad \text{a.s.}$$

                                                                                (4.56)

Proof: Grimble (1986c)

Remarks

(a)   Parts (i) and (ii) of the theorem ensure that the output and

      control signals are bounded.

(b)   If the system parameters and past values of $\varepsilon(t)$ are known the

      prediction error is obtained as 1.   Part (iii) of the theorem

      shows that this is obtained asymptotically.

(c)   From the system model shown in Figure 4.2, and from equations
(4.21) and (4.43), the transfer-function between the disturbance
$\psi_d(t)$ and the output $y(t)$ may be written:

$$y(t) = \frac{CH}{T} \psi_d(t) \tag{4.57}$$

which may be re-written in the form:

$$Ty(t) - CH\psi_d(t) = 0 \tag{4.58}$$

This equation allows the convergence result in part (iv) of the
theorem (equation (4.56)) to be more easily interpreted.

(d)   The theorem remains substantially unchanged if Assumption 4 is
replaced by a weaker assumption and the identification algorithm
is replaced by extended least-squares (see Grimble, 1986c).

4.6 PRACTICAL ISSUES

4.6.1 Control Law Implementation

When implementing the control law (4.12) it is necessary to include in the forward path the term A_r/HA_q since this term, which is common to each part of the controller (see equations (4.21), (4.24) and (4.27)), is not necessarily stable. Thus, the control law should be implemented using the equivalent structure shown in Figure 4.4 where:

$$C_{fb} \triangleq C'_{fb} \frac{A_r}{HA_q} \tag{4.59}$$

$$C_r \triangleq C'_r \frac{A_r}{HA_q} \tag{4.60}$$

$$C_{ff} \triangleq C'_{ff} \frac{A_r}{HA_q} \tag{4.61}$$

From equations (4.21), (4.24) and (4.27):

$$C'_{fb} = G \tag{4.62}$$

$$C'_r = \frac{MC}{E_r} \tag{4.63}$$

$$C'_{ff} = \frac{XC - GE_\ell D}{E_\ell A} \tag{4.64}$$

It is also necessary to show that C'_{ff} in equation (4.64) is stable. To this end, multiply equations (4.22)-(4.23) by DE_ℓ and equations (4.28)-(4.29) by C. When the resulting equations are compared the following equations, after some algebraic manipulation, are obtained:

$$\frac{CX - GE_\ell D}{A} = \frac{A_q(FDE_\ell - ZA_\ell C)}{D_c^* z^{-g}} = \frac{A_q(HDE_\ell - YC)}{BA_r} \tag{4.65}$$

By assumption the pairs A,B and A,A_r can have no unstable common factors. Since D_c^* is unstable (due to equation (4.20)) all the fractions in (4.65) are, in fact, polynomials. Thus C'_{ff} in equation (4.64) is stable (E_ℓ is stable by definition).

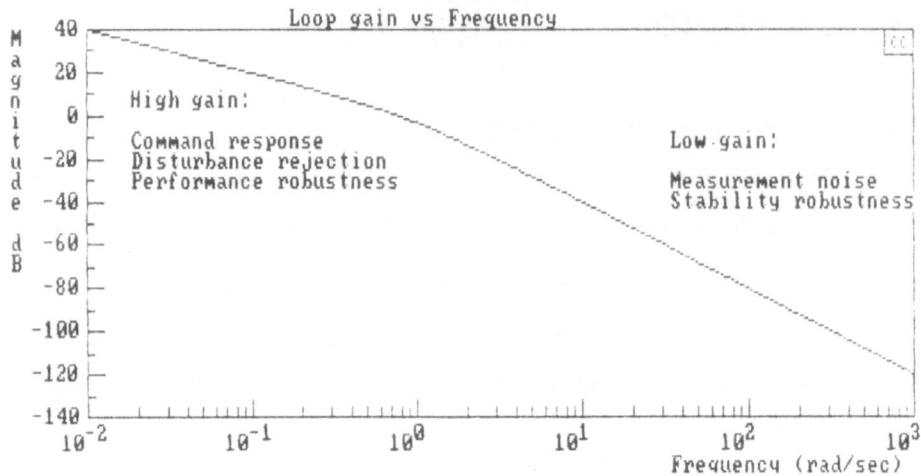

Figure 4.3 : Loop Gain

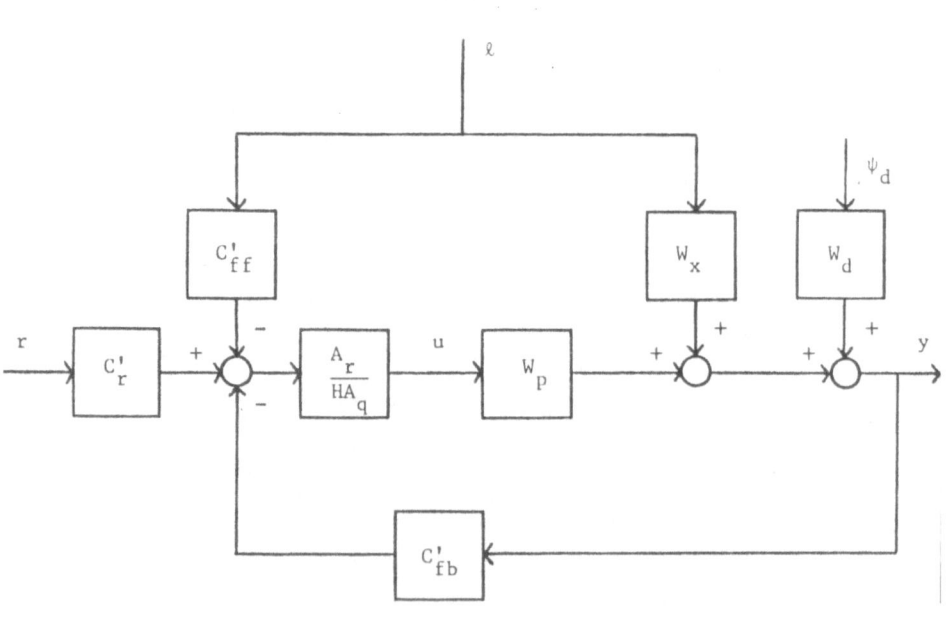

Figure 4.4 : Equivalent Control
Structure

4.6.2 Cost-function Weight Selection

In the LQG optimal controller design the cost-function weighting elements, Q_c and R_c, are the major design parameters which must be selected by the system user. Perhaps one of the key practical objectives in any self-tuning control algorithm is to reduce the number of design parameters (the 'on-line tuning knobs') to a minimum, and to give these parameters a clear physical interpretation.

In the cost-function of equation (4.17) there are many ways to choose the weighting elements, allowing various performance characteristics and loop-shaping properties to be achieved. However, a straightforward technique appropriate for self-tuning systems which involves only two design parameters, each with a simple interpretation is described in the following.

As previously mentioned, the controller will have integral action when the error weighting denominator has the form $A_q = 1 - d$. Since in the majority of practical problems it is desirable to include integral action in the controller the following definition for the cost-function weights is appropriate:

$$Q_c = \frac{B_q^* B_q}{A_q^* A_q} = \frac{(1-\beta d)^*(1-\beta d)}{(1-\beta)^2(1-d)^*(1-d)} \tag{4.66}$$

$$R_c = \frac{B_r^* B_r}{A_r^* A_r} = \frac{1}{\rho} \tag{4.67}$$

These definitions correspond to the following:

$$B_q = (1-\beta d)/(1-\beta) \tag{4.68}$$

$$A_q = 1 - d \tag{4.69}$$

$$B_r = 1 \tag{4.70}$$

$$A_r = \rho^{1/2} \tag{4.71}$$

In this formulation the scalars ρ and β effectively become the on-line tuning parameters. The interpretation of their effect is straightforward : the control weighting $1/\rho$ varies the relative magnitude of error and control penalty while β determines the amount of integral action. As ρ is increased (i.e. the control weight is decreased) the error signal will be decreased at the expense of increased control effort, an effect analogous to increasing the proportional gain of a PI controller, since the term $A_r = \rho^{1/2}$ appears in the controller numerator (see equations (4.21), (4.24) and (4.27)). As $\beta \to 1$ the integral action is removed since the term $(1-d)$ becomes cancelled in (4.66) (in practice β is never allowed to come too close to 1). Conversely, decreasing the value of β increases the effect of the integral action, which is analogous to decreasing the integral time constant in a PI controller.

Although a strict application of the theory requires that A_q is stable the use of unstable A_q, such as $A_q = 1 - d$, can be justified using the argument in the following section.

4.6.3   Solvability Conditions and Unstable Weighting Terms

Solvability of the optimal control problem is dependent upon the conditions (see Section 4.1):

1.   The polynomials A and B must have no unstable common factors. This condition is equivalent to the requirement that any unstable terms in A_d and A_x (where A_d and A_x are the denominators of W_d and W_x, respectively) must also be factors of the denominator of the plant input-output transfer-function W_p.

2.   Any unstable factors of A_e must also be factors of A.

183

3.   Any unstable factors of A_ℓ must also be factors of A and D. This condition is equivalent to the requirement that any unstable terms in $A_x A_\ell$ must also be factors of the denominator of the plant input-output transfer-function W_p.

To summarise, any unstable terms in A_e, A_d or $A_x A_\ell$ must also be factors of the denominator of the plant input-output transfer-function W_p.

In any practical situation where an unstable term in A_e, A_d or $A_x A_\ell$ does **not** appear in the denominator of W_p then this term must be artificially introduced into the forward path using A_q. Similarly, in other situations it may be desirable to use an unstable A_r weighting term. In some situations, therefore, the plant conditions may dictate that the use of unstable weighting terms is desirable.

A strict application of the theory, however, requires that A_q and A_r should be **stable**. The use of unstable A_q and A_r in practice can be justified using the following argument: let us formally define the **plant** as that part of the system which is known **a priori** in advance of controller design. Assume now that the given data is such that we know unstable weighting terms are desirable. Denote A_q and A_r as follows:

$$A_q = A_q^+ A_q^- \tag{4.72}$$
$$A_r = A_r^+ A_r^- \tag{4.73}$$

where + indicates a stable polynomial and - indicates an unstable polynomial. The forward path is then as shown in Figure 4.5. Since the given problem data tells us **a priori** that the terms A_q^- and A_r^- are necessary let us consider the transfer-function A_r^-/A_q^- as part of the plant, as shown in Figure 4.6. The controller is then designed for

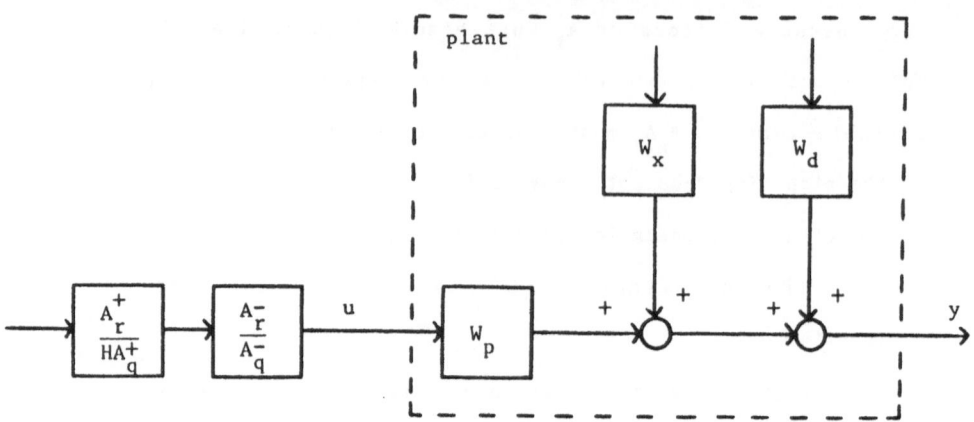

Figure 4.5 : Forward Path

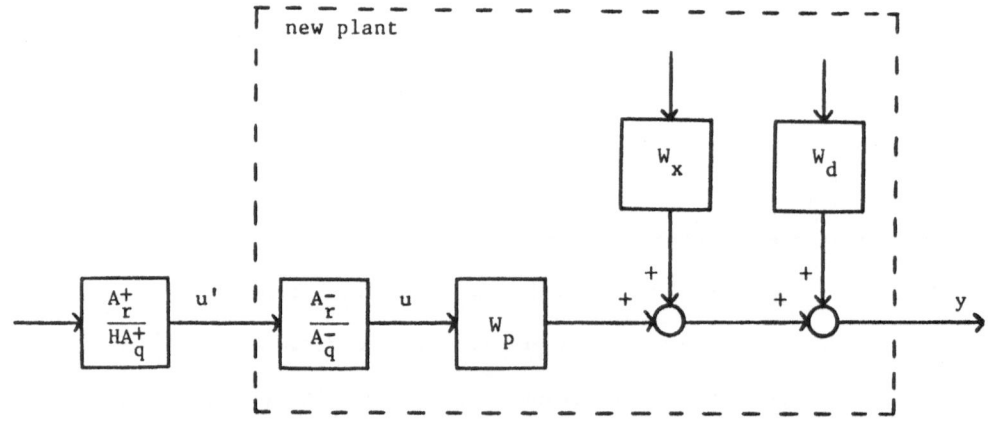

Figure 4.6 : Artificial Plant

the new plant in Figure 4.6 where the solvability conditions have now been satisfied by appropriate choice of A_q^-. This approach is equivalent to minimising the cost-function:

$$J = \frac{1}{2\pi j} \oint_{|z|=1} \left\{ \frac{B_q^* B_q}{A_q^{+*} A_q^+} \phi_e + \frac{B_r^* B_r}{A_r^{+*} A_r^+} \phi_{u'} \right\} \frac{dz}{z} \qquad (4.74)$$

Finally, the control signal u is implemented as follows:

$$u = \frac{A_r^-}{A_q^-} u' \qquad (4.75)$$

This argument allows us to ensure that the problem solvability conditions are always satisfied without violating the theoretical conditions on the cost-function weights. The original approach of directly using unstable weights will, nevertheless, result in precisely the same closed-loop system. To illustrate the point, consider a system which has a drifting disturbance due to a factor 1 − d in the denominator of the disturbance transfer-function W_d. Assume that this term is <u>not</u> present in the denominator of the plant input-output transfer-function W_p. It is immediately apparent that integral action is needed to counter the effect of the drifting disturbance. A_q is then defined as $A_q = 1 - d$ and the controller is calculated. However, that this problem violates the theoretical solvability conditions is apparent for three reasons:

(i) The polynomials A and B have an unstable common factor 1−d.

(ii) A_q is unstable.

(iii) A drifting control signal u will result in order to counteract the drifting disturbance. The cost-function will, therefore, be infinite.

The above argument can, however, be used to justify the design: let us assume that the term $1/A_q^-$ is included in the plant and then calculate a controller based on this newly defined plant to minimise the cost-function (4.74). This design will result in the same closed-loop system as the original design. The new design will, however, have the following properties.

(i)   The newly defined A and B polynomials will not have any unstable common factors.

(ii)   The effective weighting terms A_q^+ and A_r^+ are stable.

(iii) The pseudo control signal u' will be stable.

From equation (4.75) $u' = A_q^- u$. In this example $A_q^- = 1 - d$ which means that <u>changes</u> in the control signal are, effectively, being weighted.

4.6.4 Computational Algorithms

Efficient computational routines for diophantine equation solution have been derived by Kučera (1979) and Ježek (1982). The spectral factorisation can be performed using the iterative algorithms proposed by Kučera (1979) or Ježek and Kučera (1985) . Iteration of the spectral factorisation routine is terminated either when a specified tolerance is reached or when a specified maximum number of iterations have been performed. These algorithms have the necessary property that the solution obtained after each iteration is guaranteed to be stable.

4.6.5 Common Factors in A and B

The implied feedback diophantine equation (4.30) uniquely

determines the optimal feedback controller when the plant A and B polynomials are coprime. If A and B have a stable common factor then the couple of equations (4.22)-(4.23) must be solved to obtain the optimal feedback controller.

If, however, it is deemed necessary in a particular application to _always_ solve the implied equation (since this is computationally simpler than solving the original couple) regardless of any possible stable common factors in A and B then this equation will still be solvable since any stable common factors of A and B will also divide D_c (see equation (4.20)). Such a solution will lead to a closed-loop system with optimal pole positions but sub-optimal zero positions.

In this situation the common factor should be cancelled from both sides of the equation before it is solved. The algorithm derived by Jezek (1982) is based upon the Euclidean algorithm which can inherently cope with such common factors.

This property of the optimal design should be contrasted with standard pole-assignment algorithms (Wellstead et al, 1979) where the diophantine equation (4.44) must be solved. Since the arbitrary polynomial T_c appears on the right-hand-side of this equation (in place of D_c) _any_ common factors in A and B will render this equation unsolvable (unless, coincidentally, this factor also divides T_cC).

188

Example 4.1

In this example the technique outlined in Section 4.6.3 for dealing with unstable reference generator poles which are not also poles of the plant is illustrated. Consider a sinusoidal reference of the form $r(t) = \sin \omega kT$, where k is the sample instant and T the sample period. When $\omega = \pi$ and $T = 0.1$, this reference signal may be modelled as:

$$W_r = A_e^{-1}E_r = \frac{0.31}{1-1.9d+d^2}$$

Consider a plant defined by:

$$W_p = \frac{d^2(1 + 0.5d)}{1-0.95d}$$

To formulate a meaningful optimisation problem the technique outlined in Section 4.6.3 is used : A_q is defined as $A_q = A_q^+A_q^-$, where $A_q^+ = 1$ and $A_q^- = 1-1.9d + d^2$, the unstable poles of W_r. The term $1/A_q^-$ is now considered part of the plant, so that the effective plant becomes:

$$W_p' = A^{-1}B = \frac{d^2(1 + 0.5d)}{(1-0.95d)(1-1.9d+d^2)}$$

The remaining cost function weights are defined simply as $B_q = A_r = B_r = 1$. From equations (4.20)-(4.26) the optimal feedback and reference controllers may be calculated (replacing A_q by A_q^+) as:

$$C_{fb} = \frac{7.44 - 10.93d + 4.46d^2}{3.02 + 5.87d + 2.35d^2}$$

$$C_r = \frac{2.98 - 2.46d}{3.02 + 5.87d + 2.35d^2}$$

The resulting closed-loop system was then simulated and a sinusoidal set-point applied. The plant output and set-point are shown in Figure 4.7(a). The control signals $u(t)$ and $u'(t)$ are shown in

Figure 4.7(b). It is seen that perfect tracking is achieved, that u(t) is oscillatory, and that the steady-state value of u'(t) is zero.

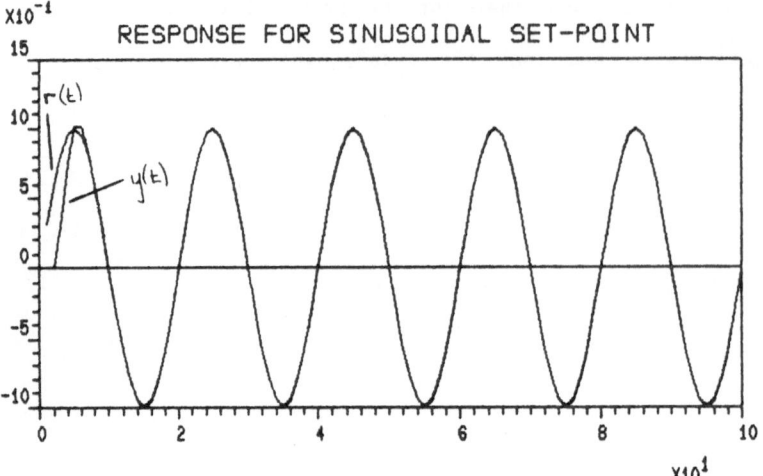

Figure 4.7(a)

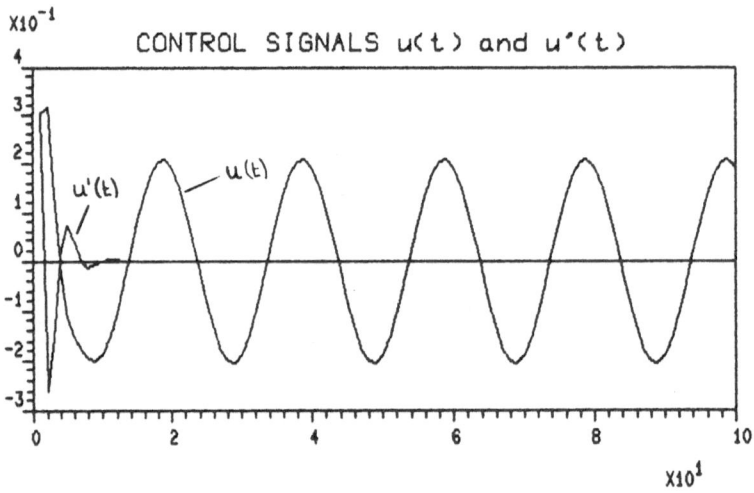

Figure 4.7(b)

Example 4.2

This example is a self-tuning version of example 4.1. The plant parameters were estimated during each sample interval using Recursive Least-squares (RLS). For simplicity a constant forgetting factor of unity was used. The plant output and set-point are shown in Figure 4.8(a). The control signals $u(t)$ and $u'(t)$ are shown in Figure 4.8(b). After an initial tuning-in transient the plant output again follows the set-point.

Estimates of the plant A and B polynomials are plotted in Figure 4.9. Evolution of the resulting controller parameters is shown in Figure 4.10. It is seen from Figure 4.10 that the controller parameters converge to the values calculated in Example 4.1 using the true plant polynomials.

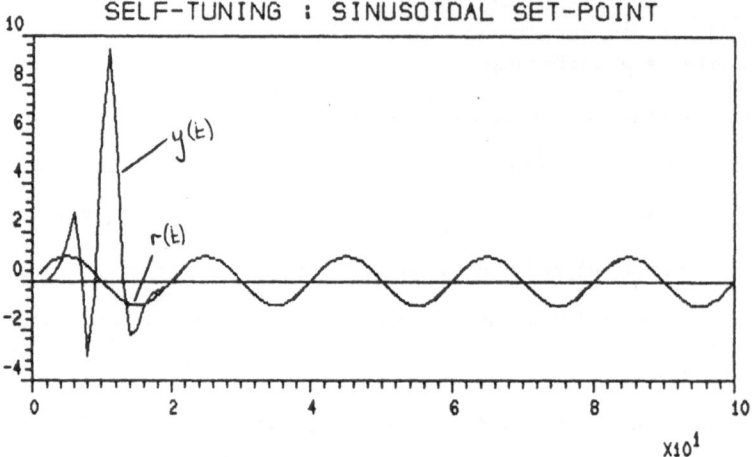

Figure 4.8(a)

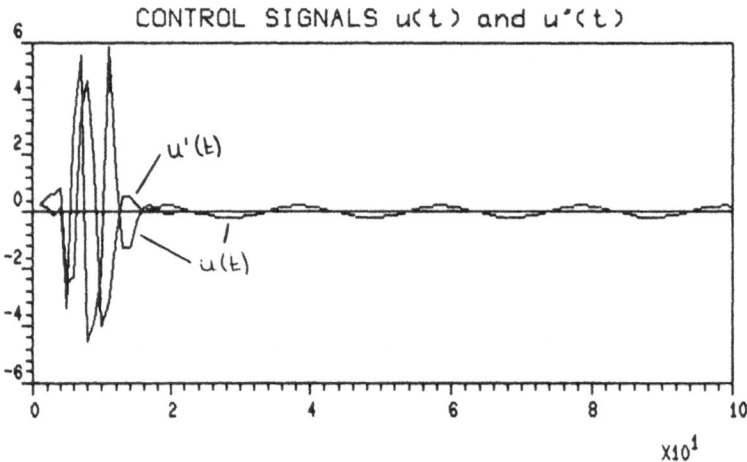

Figure 4.8(b)

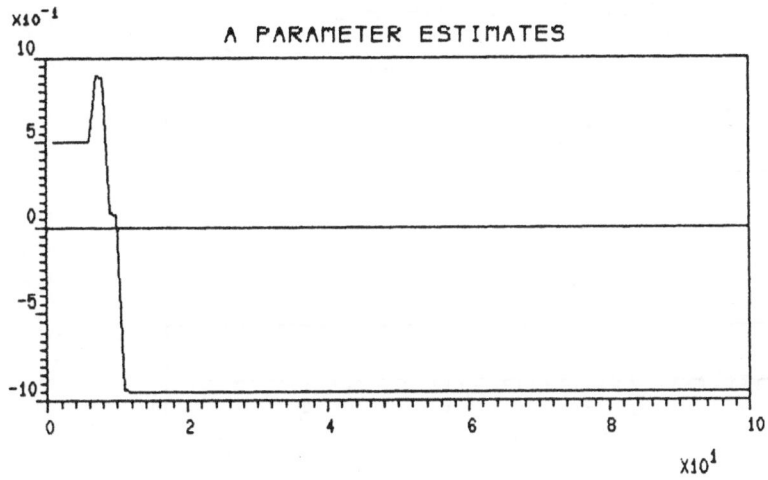

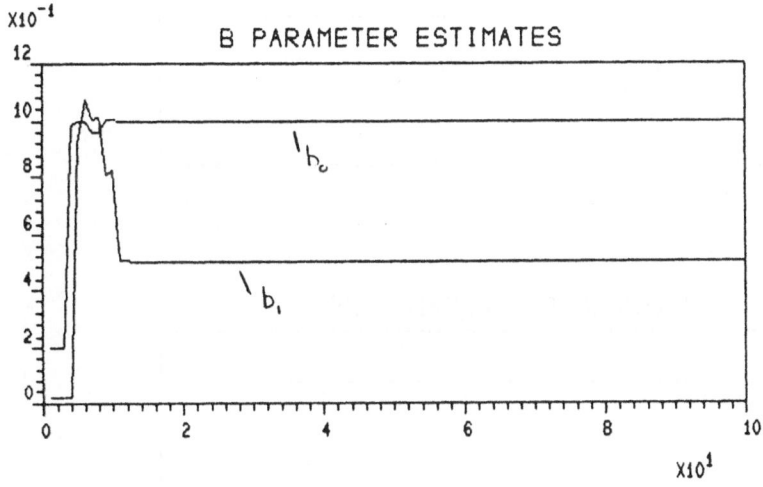

Figure 4.9

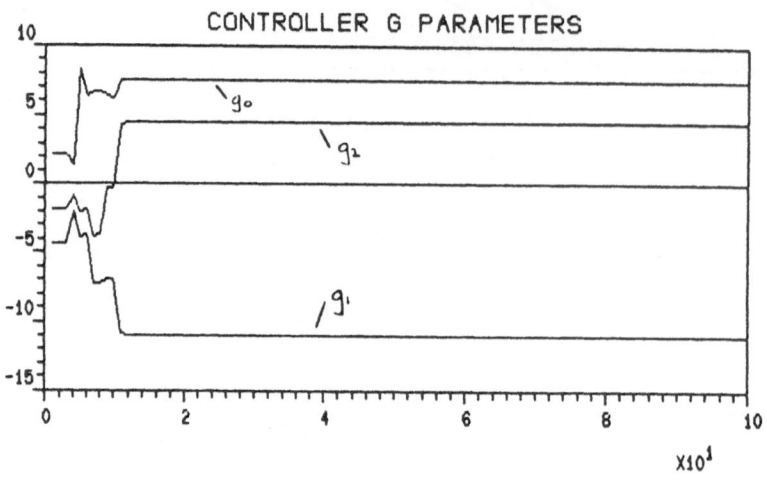

CONTROLLER G PARAMETERS

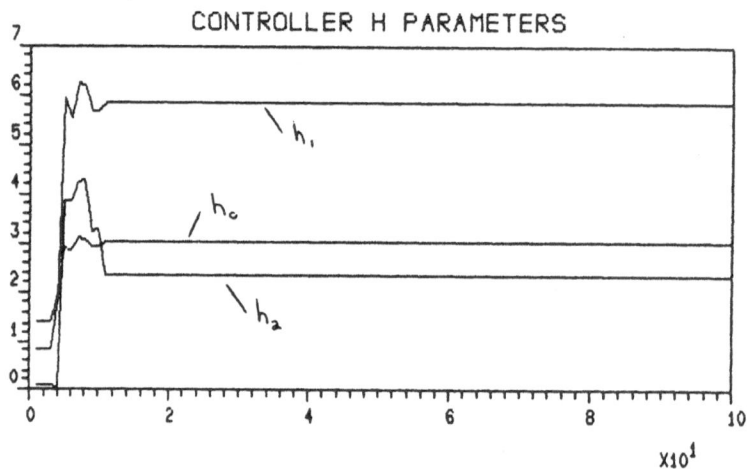

CONTROLLER H PARAMETERS

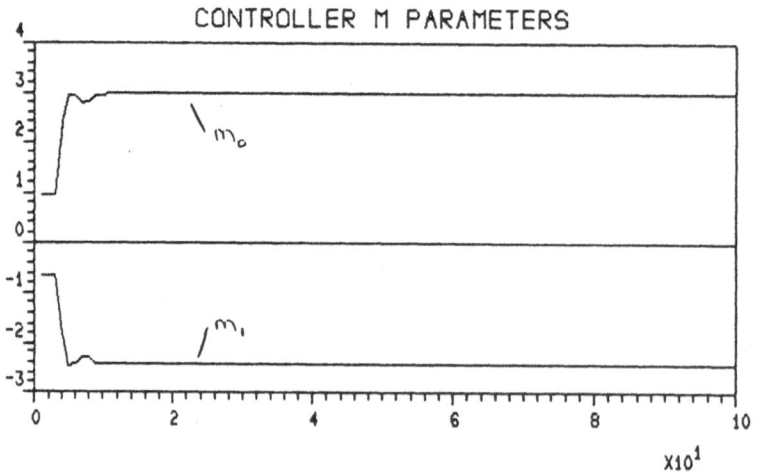

CONTROLLER M PARAMETERS

Example 4.3

This example investigates the tracking and feedforward control performance in the self-tuning context. Consider the following plant:

$$W_p = A^{-1}B = \frac{d^2(1+ 0.5d)}{1-0.95d}$$

$$W_d = A^{-1}C = \frac{1}{1-0.95d}$$

$$W_x = A^{-1}D = \frac{d^2(1-0.75d)}{1-0.95d}$$

The plant output was required to track step-like changes in set-point while the plant was subjected to step changes in the measurable load disturbance $\ell(t)$. The reference and load disturbance generators were therefore modelled as:

$$W_r = A_e^{-1}E_r = \frac{1}{1-d}$$

$$W_\ell = A_\ell^{-1}E_\ell = \frac{1}{1-d}$$

Integral action was included in the controller by defining $A_q = 1-d$. The control weighting R_c was set to $R_c = 0.1$, and B_q was chosen as $B_q = 1$. Using the above data the <u>true</u> controller transfer functions may be calculated as:

$$C_{fb} = \frac{3.01 - 2.01d}{(1.35+2.71d+1.06d^2)(1-d)}$$

$$C_r = \frac{1}{(1.35+2.71d+1.06d^2)(1-d)}$$

$$C_{ff} = \frac{1.04+0.20d - 2.82d^2 +1.59d^3}{(1.35+2.71d+1.06d^2)(1-d)}$$

In the simulation results which follow the load disturbance $\ell(t)$ changed as follows:

$t < 60$ $\quad\quad\quad \ell(t) = 0$

$60 < t < 90$ $\quad\quad \ell(t) = 10$

$90 < t < 120$ $\quad\quad \ell(t) = 0$

$120 < t < 150$ $\quad\quad \ell(t) = 10$

$t > 150$ $\quad\quad\quad \ell(t) = 0$

Performance of the fixed controller with C_{fb} and C_r as above but with no feedforward action ($C_{ff} = 0$) is shown in Figure 4.11. The effect of the changes in load disturbance is clearly seen, with the disturbance eventually being rejected after each change by the integral action only.

Performance of the fixed controller including optimal feedforward control is shown in Figure 4.12. The effect of the load disturbance is greatly reduced in this case. Note that a further reduction in the effect of $\ell(t)$ could be achieved by reducing the control weighting (in fact, since the delay in D = delay in B complete cancellation is possible for $R_c = 0$).

The performance for the self-tuning simulation is shown in Figure 4.13. Again, RLS with unity forgetting factor was used to estimate the A,B and D polynomials. The first change in load disturbance is clearly seen on the output. However, after the feedforward controller has tuned-in the performance matches that of the fixed controller shown in Figure 4.12. The control signal for the self-tuning simulation is also shown in Figure 4.13.

Estimates of the plant polynomials A,B and D are shown in Figure 4.14.

Evolution of the feedback controller G and H polynomials is shown in Figure 4.15, and the feedforward controller numerator C_{ffn} in Figure 4.16. Comparison with the fixed controllers calculated above shows that the parameters converge to their true values.

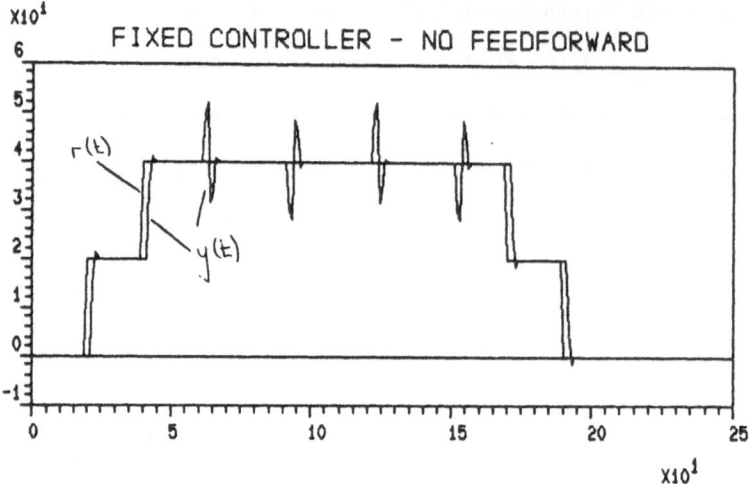

Figure 4.11

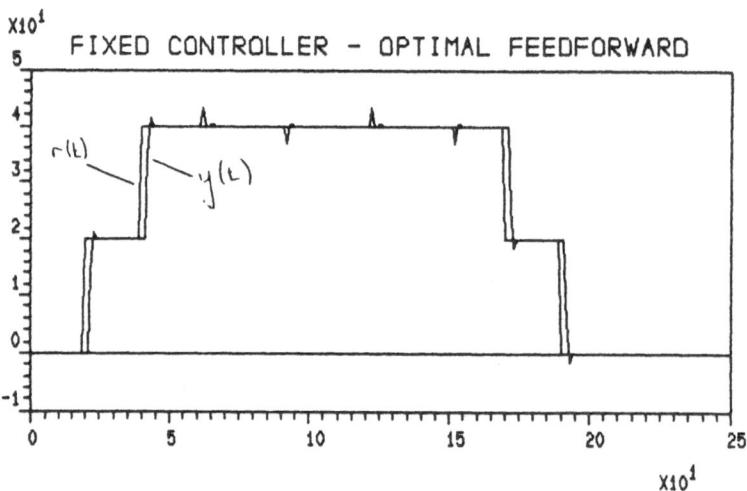

Figure 4.12

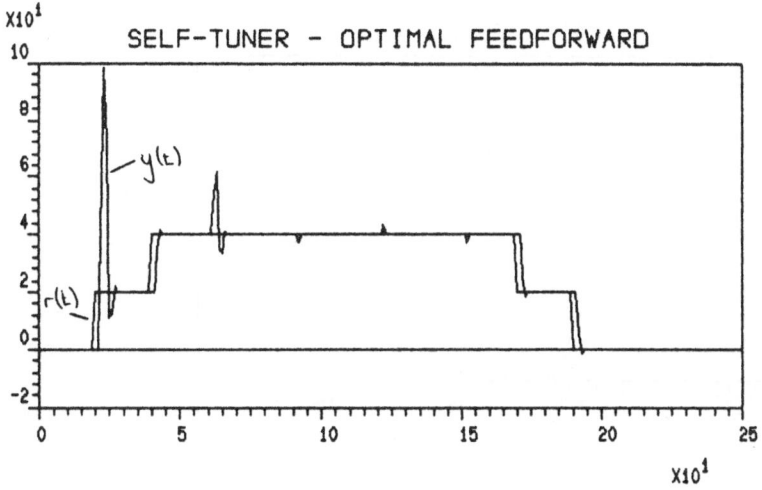

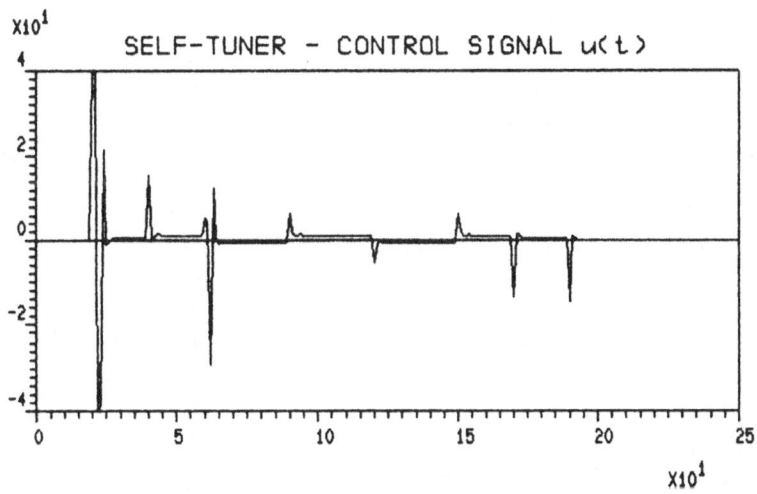

Figure 4.13

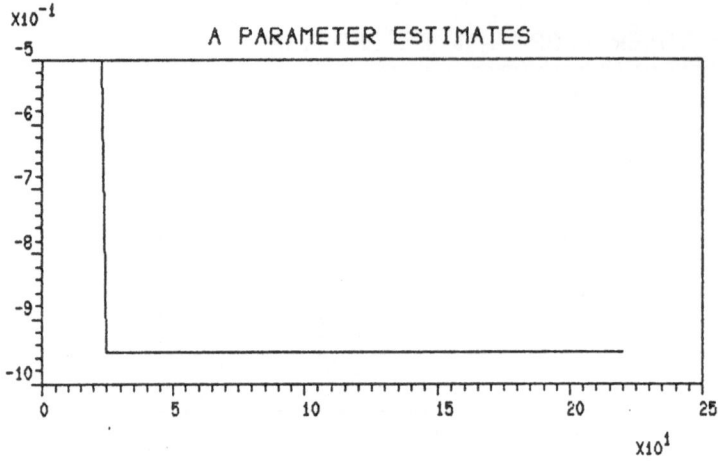

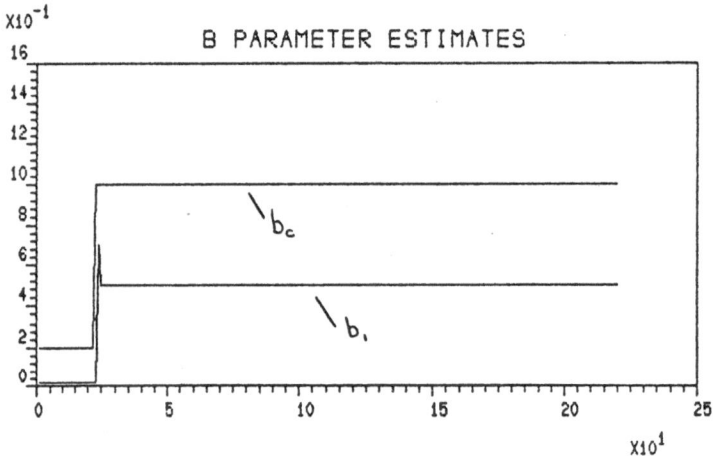

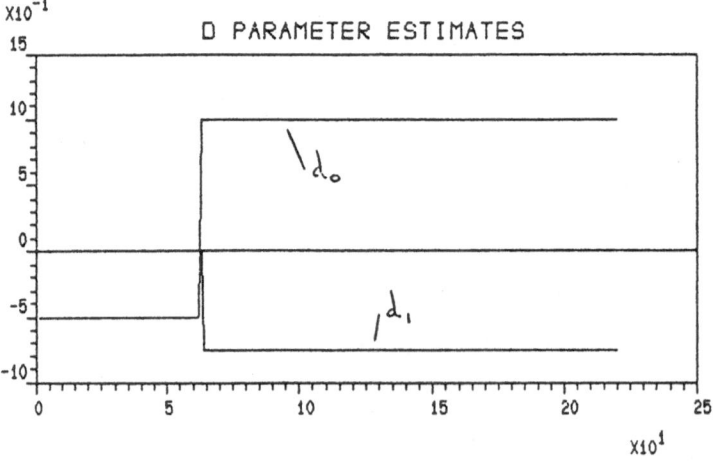

Figure 4.14

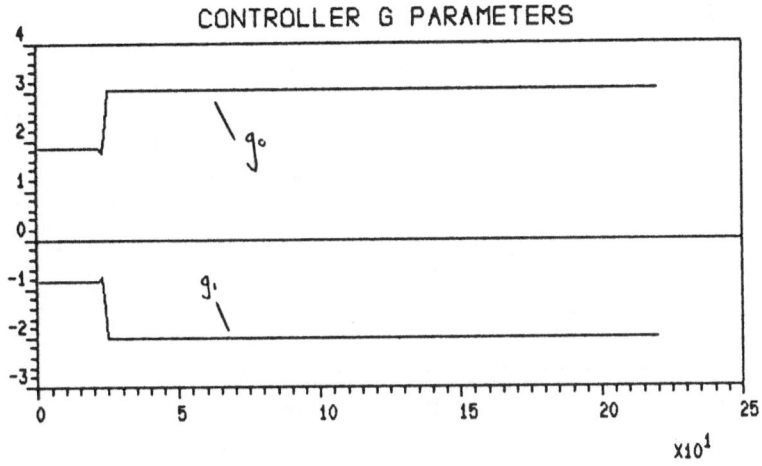

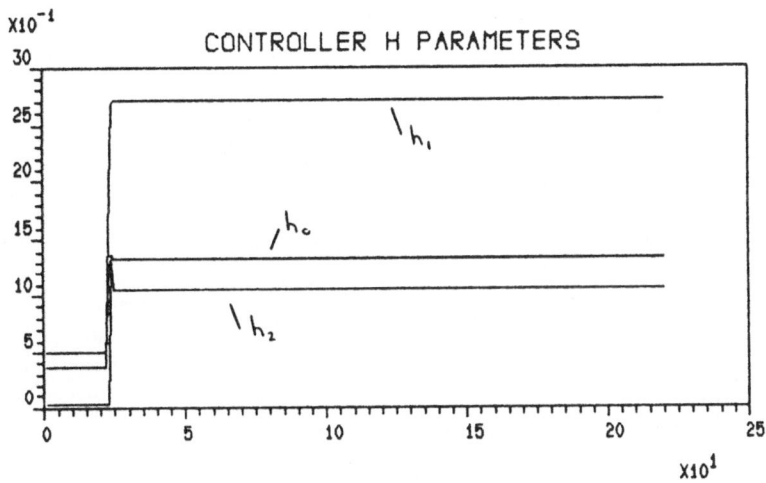

Figure 4.15

202

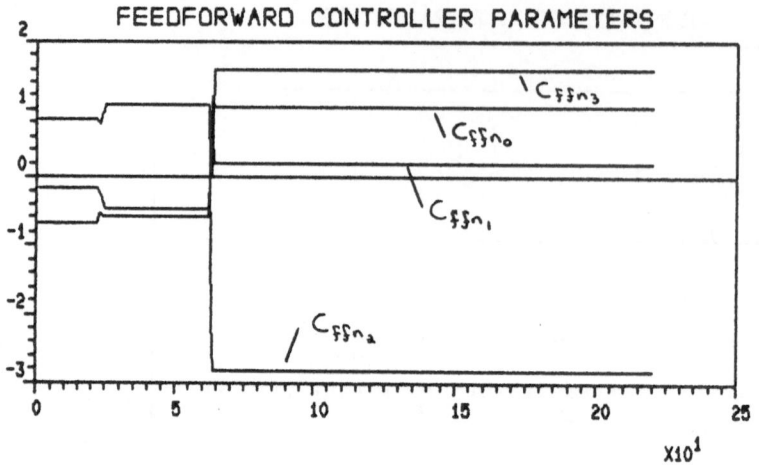

Figure 4.16

PART THREE

CASE STUDY

CHAPTER FIVE

A POWER SYSTEMS APPLICATION

Summary

The self-tuning LQG controller described in the previous chapter has been evaluated in experimental trials at the Hunterston 'B' Advanced Gas-cooled Reactor (AGR) power station simulator. The results of these trials are reported in this chapter.

The principles of the AGR power system design are briefly outlined in Section 5.1. The control strategy currently implemented at Hunterston is described in Section 5.2. The Hunterston 'B' simulator consists of a full-scale replica of the station central control room and the plant models run in real-time on a distributed network of 52 processors. The simulator is described in Section 5.3.

The control loop on which the LQG self-tuner was implemented is the turbine stop-valve (TSV) steam pressure loop. The existing controller on this loop is of analogue Proportional-Integral (PI) form. Performance of the PI controller is evaluated in Section 5.4. The results of the self-tuning LQG trials and details of the hardware/software implementation of the algorithms are presented in Section 5.5.

The chapter concludes in Section 5.6 with an evaluation of the LQG controller performance, and a comparison with the performance of the existing PI controller.

5.1   THE ADVANCED GAS-COOLED REACTOR

The Hunterston 'B' nuclear power station is based upon the
Advanced Gas-cooled Reactor (AGR) design (see Figure 5.1). The fuel
elements used in the AGR consist of uranium oxide encased in
stainless steel cladding to form fuel rods. The fuel rods are loaded
into vertical channels in the reactor core, which is made up of
graphite bricks (the moderator).   The core has further vertical
channels which contain control rods.   The control rods are made of
strong neutron absorbing material and can be inserted or withdrawn
from the core to adjust the rate of the fission process and hence the
reactor heat output.

The graphite moderator and fuel elements are cooled by
circulating pressurised carbon dioxide gas.   The gas temperature is
thereby raised and the hot gas is then passed to the boilers to
produce steam which subsequently drives the turbines.

The whole assembly is encased in a pre-stressed concrete
pressure vessel which performs the dual purpose of gas containment
and radiation shielding.

For reliability reasons the boiler/reactor assembly is divided
into four distinct quadrants.   Each quadrant consists of three
boilers and two gas circulators, and the boilers are further divided
into half-units which means that each reactor consists of a total of
twenty-four water-steam circuits.

5.2   THE HUNTERSTON 'B' OVERALL CONTROL STRATEGY

The overall reactor/turbine generation system consists of two
closed circuits.   In the primary circuit pressurised carbon dioxide
gas is pumped up through the reactor core and over onto the boiler

207

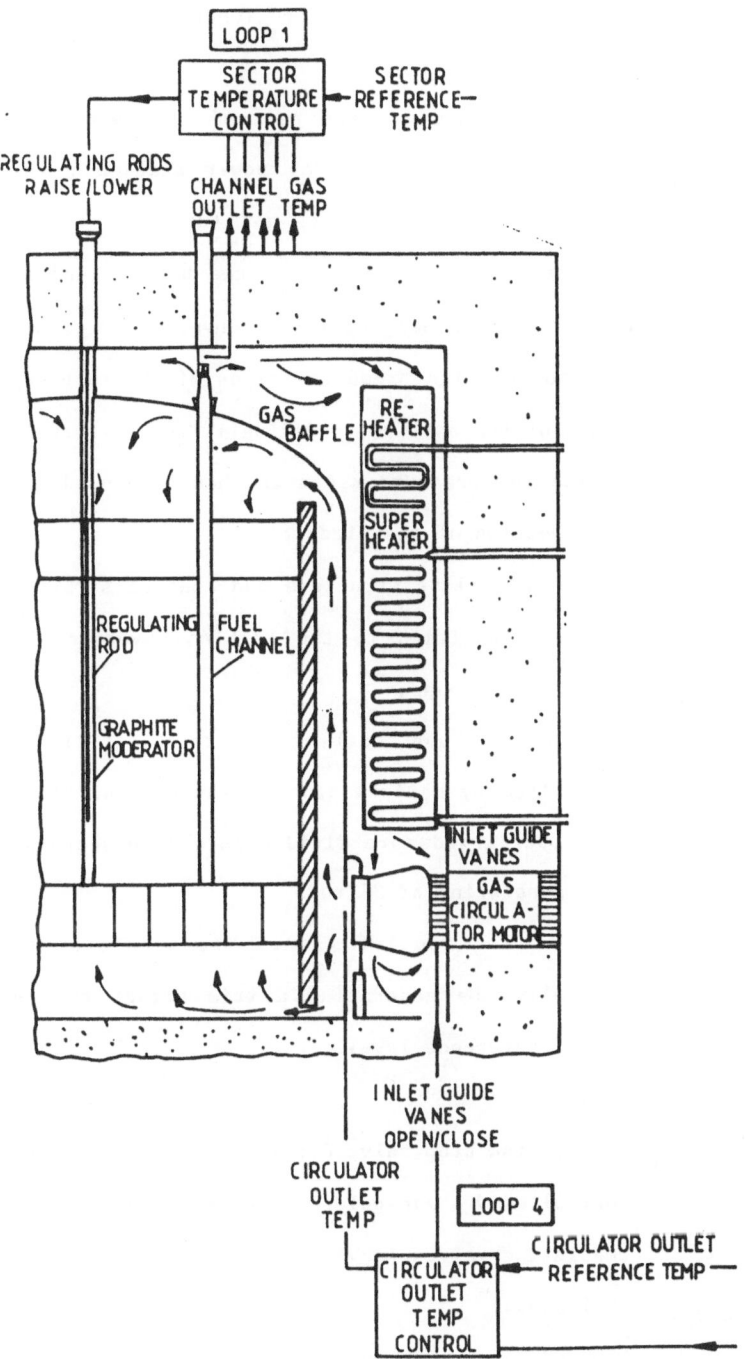

Figure 5.1 : The Advanced Gas - Cooled
Reactor

heat exchangers by the gas circulators which are located beneath the boilers.

Having generated steam in this way the rest of the generation plant is similar in nature to a conventional power station ; the secondary circuit is comprised of the boilers and turbine. In the secondary circuit feed pumps supply feedwater to the boilers and as it is passed up through the heat exchangers it is converted to steam. The steam is then passed to the high pressure stages of the turbine, back to the reheater banks of the boilers, and finally to the turbine low pressure stages. The low pressure steam is then condensed and passed back into the feed section of the circuit.

A schematic of the overall plant structure and control strategy is shown in Figure 5.2. In addition to the turbine governor the system consists of seven control loops:

(1)   Loop 1: The reactor gas outlet temperature (T2) is controlled by manipulation of the control rods. The control rods are divided into five sectors, each consisting of 37 rods.

(2)   Loop 2: The boiler outlet steam temperature is controlled by generating a trim signal to the Loop 1 reference level.

(3)   Loop 3 (main):

The turbine stop valve (TSV) steam pressure is controlled by manipulation of the boiler feed regulating valves (FRV's).

(4)   Loop 3 (auxiliary):

The differential pressure across the feed regulating valves is controlled by manipulation of the feed pump speed.

209

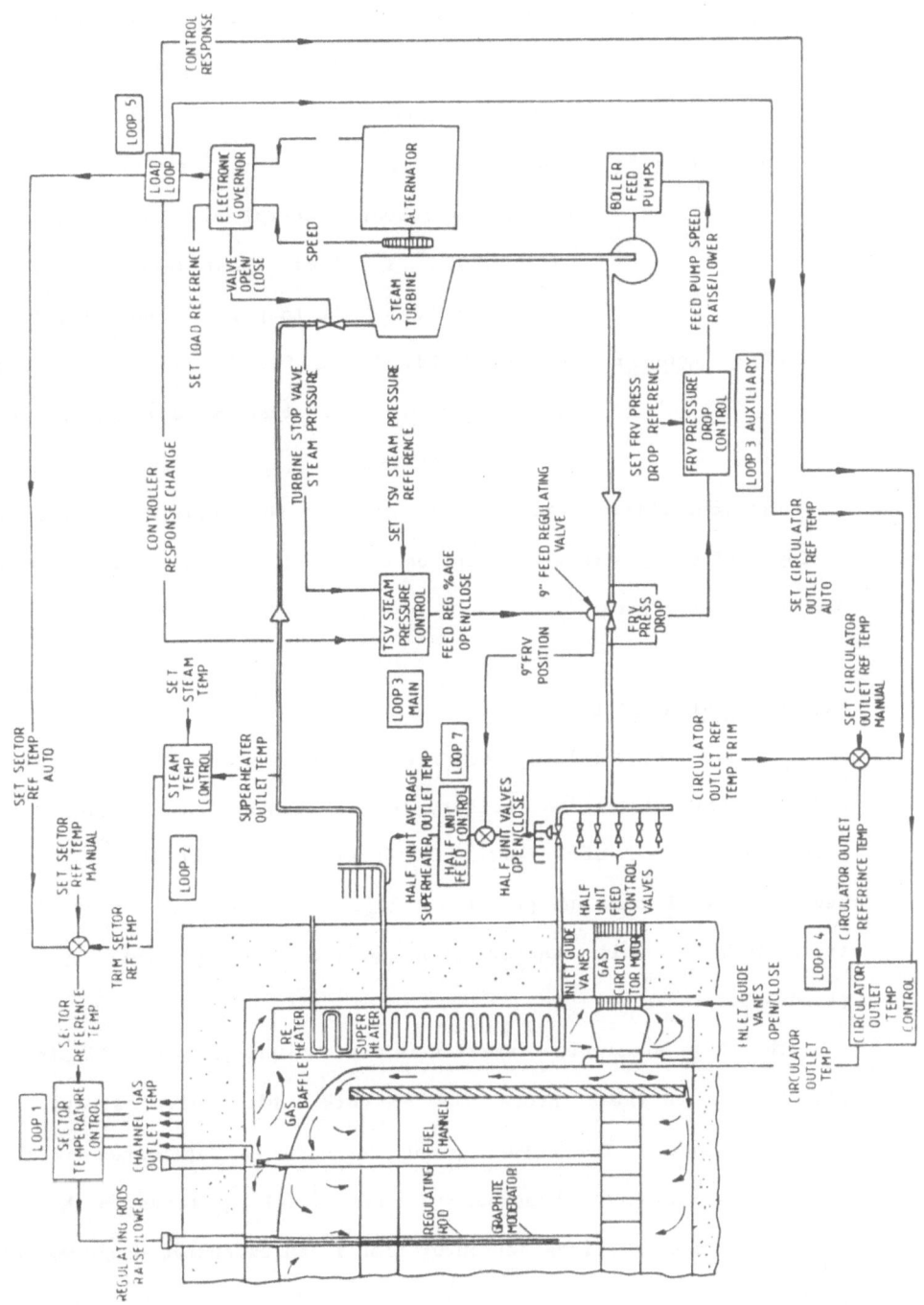

Figure 5.2 : Hunterston 'B' Overall
Control Strategy

(5)   Loop 4: The circulator outlet gas temperature (T1) is controlled by manipulation of the gas circulator inlet guide vanes.

(6)   Loop 5: Changes in demanded turbine load are used to provide feedforward control for Loop 3 (main) and Loop 4.   The load signal is also used to provide a trim for the Loop 1 reference level.

(7)   Loop 7: The individual half-unit outlet steam temperatures are controlled by manipulation of each half-unit valve.

The controllers which perform each of the above tasks are of analogue PI form, with the exception of Loop 1 controller which is of double lead-lag form.

5.3   HUNTERSTON 'B' SIMULATOR

The South of Scotland Electricity Board have built at Hunterston a total plant simulation facility for the Hunterston 'B' AGR.   The simulator consists of an exact full-scale replica of the station's central control room and provides full simulation of all major plant items including all reactor plant protection and safety systems (Figure 5.3).

The whole simulation facility is run in real-time by utilising a distributed parallel processing network which uses arrays of microprocessors.   The actual plant response is duplicated by the real-time solution of thousands of differential equations which have been obtained over a long period of time in an extensive programme of plant modelling studies.   By duplicating all instruments, controls

<u>Figure 5.3</u> : Simulator Control Room

and displays in the control room and connecting these to the
real-time simulation an environment is created which is
indistinguishable from the real thing.

The main purpose of the simulator is to provide a full training
facility for all control room personnel.   As well as providing the
opportunity for the rehearsal of routine plant operations such as
start-up and shutdown the simulator is also used to investigate a
wide variety of fault conditions.   The simulator is monitored from
an instructors console which allows the plant to be initialised at
any given state and subsequently to be subjected to any desired
sequence of operating conditions.

In addition, the simulator provides the ideal environment for
the testing and evaluation of new control methods and techniques
which aim to modify and improve the efficiency of the existing
strategy.

The Hunterston 'B' models are mounted on a total of 52 Marconi
computers.   These 52 processors are split into several clusters
which each simulate various areas of the plant.   Each cluster is
connected in parallel to the central control cluster.   The computers
are distributed throughout the control room and drive the display
panels via interface equipment.

The adaptive LQG controller which was tested at the simulator
was mounted in an IBM PC in which an I/O card was installed (see
Section 5.5.1 for more details).   Connection of the IBM to the
simulator involved small modifications to the simulator software in
order to remove the existing PI controller from the loop under
investigation.   The IBM controller was then introduced into the loop
by directly connecting via its I/O card to the simulator interface

equipment behind the display panels (see Figures 5.4 and 5.5).

5.4 TSVP CONTROL (LOOP 3 (MAIN))

The control loop under investigation in this particular study is
the turbine stop valve steam pressure (TSVP) control, which is
achieved by manipulation of the feed regulating valves (Loop 3
(main)).   This loop is of particular importance in the overall
system since the maintenance of a steady TSV pressure has a direct
influence on the power output of the generator. In addition, since
Loop 3 (main) regulates the secondary water-steam circuit it has a
direct influence on the other controlled variables in the system.
Thus, the study investigates the following factors:

(i)   The transient response of Loop 3 (main).

(ii)   The tightness of control in steady-state.

(iii)   The transients induced into the rest of the system due to
perturbations in the Loop 3 (main) reference level.

A primary objective of the study is to investigate the
performance of the LQG self-tuner with respect to the above factors.
To evaluate the LQG controller the performance of the existing
analogue PI controller is first studied.

5.4.1 PI Performance

The TSV steam pressure reference level was subjected to two step
changes : a step from 85% to 65% followed by a step from 65% back to
85% (the pressure scaling was chosen such that 0-100% corresponds to
120-170 bar). The interaction between Loop 3 (main) and the rest of
the system was investigated by monitoring three other controlled

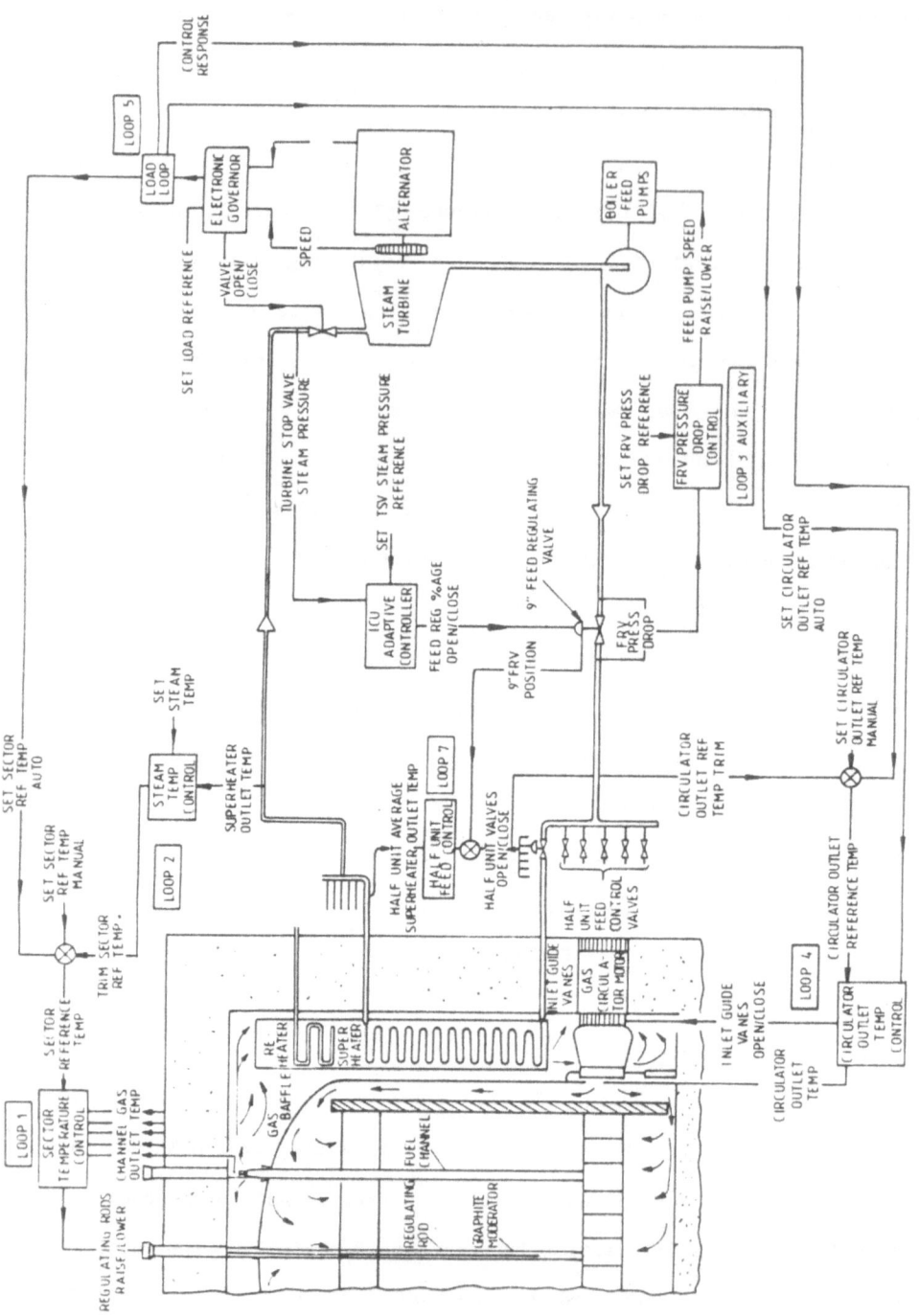

Figure 5.4 : Overall Control Scheme with
IBM Adaptive Controller

Figure 5.5 : IBM Adaptive Controller

variables : the circulator outlet gas temperature (T1), the reactor gas outlet temperature (T2), and the feed regulating valve differential pressure (FRVDP).

The responses for the downward step are shown in Figure 5.6 and for the upward step in Figure 5.7. Each figure has three graphs:

(i)   TSV pressure and set-point.

(ii)   FRV position.

(iii)   A plot of the other monitored variables T1, T2 and FRVDP (along with TSVP and FRV position).

From Figures 5.6 and 5.7 the following observations regarding the performance of the PI controller may be made:

(i)   Each change in the TSV steam pressure reference level is followed by a sudden and large movement of the FRV actuator. This initial valve movement is followed by a slow oscillatory transition to the steady-state region.

(ii)   As a result of (i) the TSV steam pressure response is initially quite fast but displays a significant overshoot and oscillation around the reference level when moving into the steady-state region.

(iii)   The initial rapid movement of the FRV actuator induces very strong transients in the other controlled variables T1, T2 and FRVDP. The FRVDP transient is particularly severe. Following the initial period the transients decay with a slow oscillation.

217

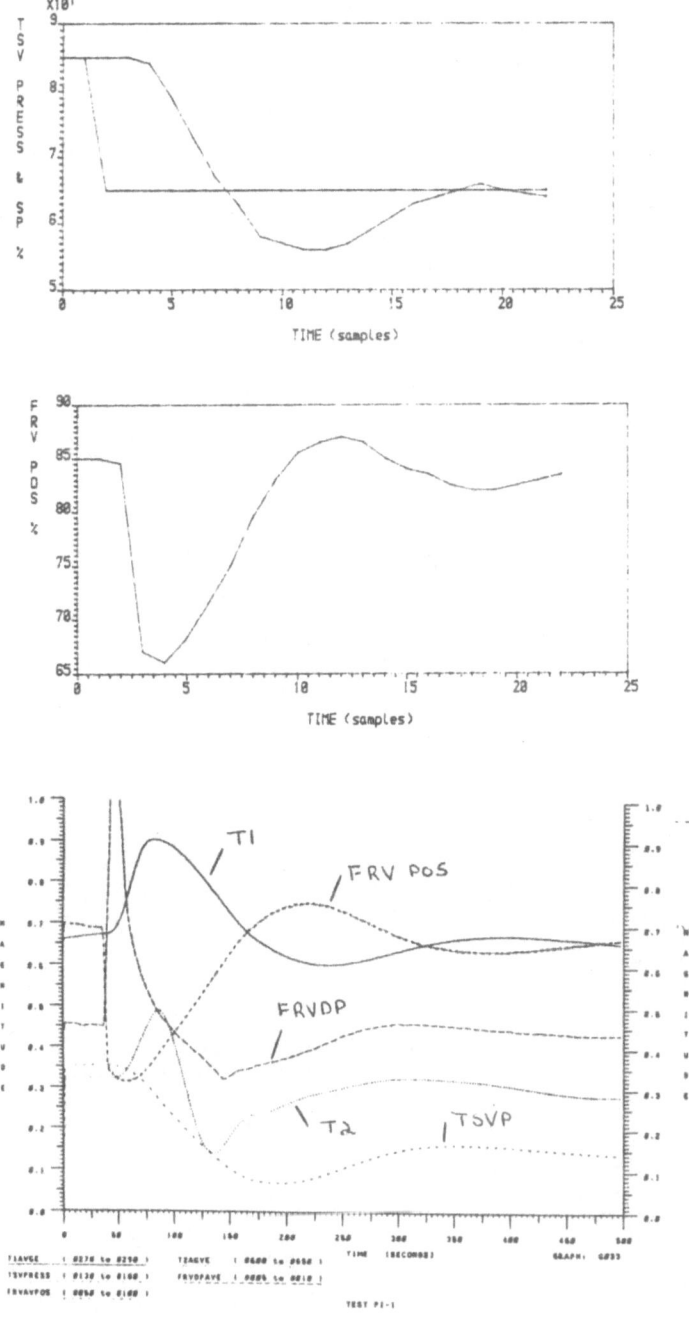

Figure 5.6 : PI Control

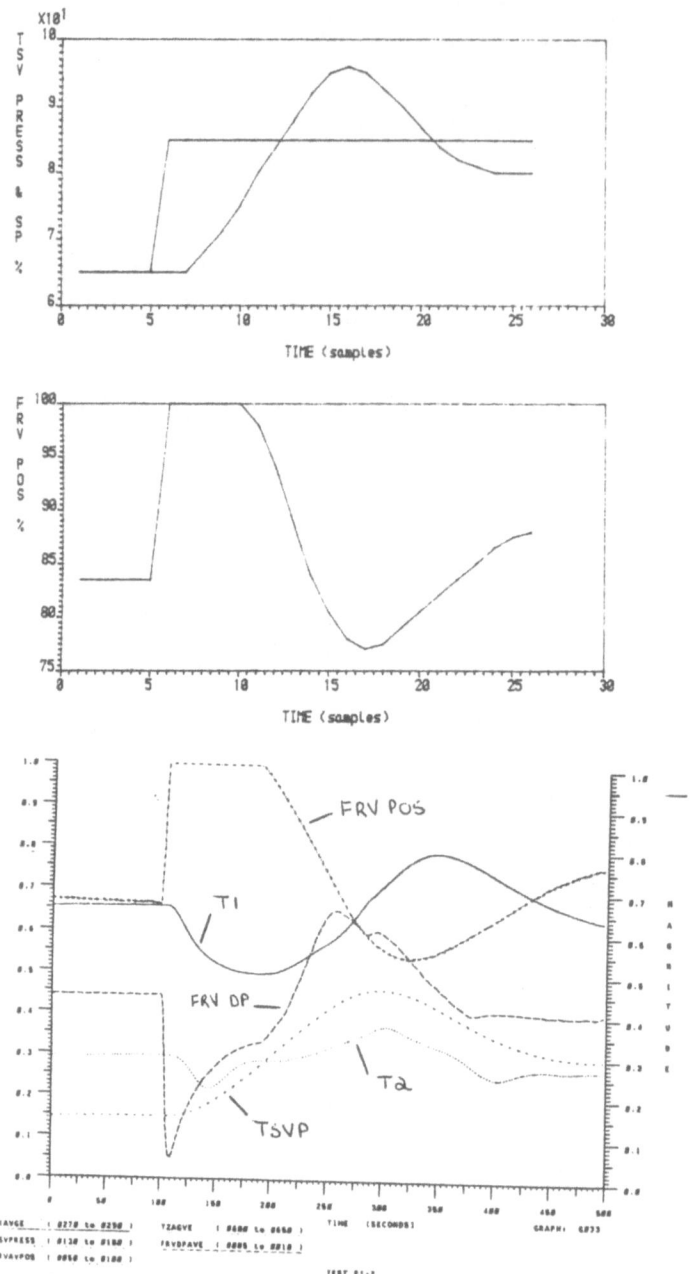

Figure 5.7 : PI Control

5.5 SELF-TUNING LQG CONTROL

5.5.1 Hardware and software

The LQG self-tuner was implemented on an IBM PC in which an IBM Data Acquisition and Control Adapter (DACA) card was installed. Interface to the card was via an external terminal board (see Figure 5.5).

The real-time LQG self-tuning algorithm was programmed mainly in IBM Professional FORTRAN (FORTRAN 77). All graphics were drawn using the IBM Plotting System and data I/O was performed by calls to the DACA subroutine library. A keyscan routine for real-time operator interface was programmed in 8087 assembly language.

A fuller technical summary of the PC-based self-tuning system is given in Hunt and Jones (1988).

5.5.2 Control law and model parameterisation

For this particular study the plant model used was given by:

$$y(t) = \frac{B}{A} u(t) + \frac{C}{A} \psi_d(t) \tag{5.1}$$

The control law used is given by:

$$u(t) = -C_{fb}y(t) + C_r r(t) \tag{5.2}$$

This controller structure and plant model corresponds to the design presented in Chapter 4, but without the feedforward part of the controller.

For the cost-function (4.17) the weighting functions Q_c and R_c were chosen according to the formulation given in Section 4.6.2 i.e:

$$Q_c = \frac{(1-\beta d)^*(1-\beta d)}{(1-\beta)^2(1-d)^*(1-d)} \quad , \quad R_c = \frac{1}{\rho} = \mu$$

In the following trials the scalar β was fixed at $\beta = 0.1$ so that the only on-line tuning parameter was the control weighting μ.

In the controller design stage the feedback part of the controller was calculated using the implied feedback diophantine equation (4.30) and the reference part was calculated using the steady-state strategy defined by equations (4.36) and (4.38). The control design steps may be summarised as follows:

(i)   Calculate D_c from the spectral factorisation (4.20) i.e.:

$$D_c^* D_c = B^*(1.11-0.11d)^*(1.11-0.11d)B + A^*(1-d)^*\mu(1-d)A$$

(ii)   The feedback controller is given by:

$$C_{fb} = \frac{G}{H(1-d)}$$

where G,H are calculated from the diophantine equation (4.30):

$$A(1-d)H + BG = D_c C$$

such that:

$$\deg G = \deg A$$

(iii)   The reference controller is given by equation (4.36):

$$C_r = \frac{\gamma C}{H(1-d)}$$

where γ is calculated according to equation (4.38) as:

$$\gamma = \frac{D_c(1)}{B(1)}$$

The control law is then implemented according to the strategy given in Section 4.6.1.

Based on the results of an open-loop step test performed on Loop 3 (main) a sample interval of 20 seconds was chosen. For the parameterisation of the estimation routine a second order model with a two-step delay was selected such that the estimated A and B polynomials were given by:

$$A = 1 + a_1 d + a_2 d^2$$
$$B = d^2(b_0 + b_1 d)$$

The C polynomial was simply set to C = 1 so that a total of four parameters were estimated. The estimation routine used was recursive least-squares with an ordinary exponential forgetting factor which could be altered on-line from the keyboard.

The parameter estimates were initialised according to the following stable second order model:

$$\frac{B}{A} = \frac{d^2(0.15 + 0.05d)}{1- 1.5d + 0.7d^2}$$

and the forgetting factor was initially set to $\lambda = 0.95$.

5.5.3 Self-tuning LQG control results

The performance of the LQG self-tuner was investigated by performing two main trials. In the first trial the main objective was simply to obtain accurate estimates of the plant parameters, and subsequently to investigate the effect of varying the control weighting while the estimation routine was frozen. In the second trial the estimation routine was frozen for the whole run and the loop subjected to a series of step changes in the set-point while the control weighting was varied. The step changes were of the same magnitude as those applied during the PI test described in Section 5.4.1. During each step change the responses of the other controlled variables in the system (T1, T2 and FRVDP) were recorded in order that the LQG controller could be compared with the existing analogue PI:

(i)   Run 1

The first test had a duration of 260 samples (1 1/2 hours). The TSV steam pressure and set-point, and the FRV position are shown in Figure 5.8. The parameter

estimation was frozen at t = 120. During the tuning
phase (t < 120) the loop was subjected to a sequence of
step-changes in set-point. The loop responses for the
tuning phase are shown in Figure 5.9. Evolution of the
parameter estimates and prediction error during this time
is shown in Figure 5.10.

The step response at t = 100 indicated that the
parameter estimates were of reasonable accuracy and this
led to the decision to freeze estimation at t = 120. At
this time the estimated plant model was:

$$\frac{B}{A} = \frac{d^2(0.001 + 0.08d)}{1 - 1.158d + 0.2548d^2}$$

The low relative value of the first B coefficient
indicates that the plant time-delay may have been
underestimated by one step.

A relatively low value of control weighting,
μ = 0.8, was used during the tuning phase. After the
parameter estimation routine was frozen the control
weighting was successively increased during the fixed
phase (t > 120). The loop responses for the fixed phase
are shown in Figure 5.11. Changes in control weighting
occurred as follows:

t = 145 , μ = 2.0

t = 160 , μ = 10.0

t = 170 , μ = 30.0

t = 180 , μ = 40.0

As would be expected, the graph of FRV position shows

223

that the effect of increasing the control weighting is to significantly damp out the actuator movements.

Comparison of the step response occurring at t = 100 (μ = 0.8) with those at t = 190 and t = 230 (μ = 40.0) also shows, however, that increasing μ also leads to a slower step response with significant overshoot.

The results clearly demonstrate that the particular formulation of the control weights used allows the actuator activity to be traded against closeness of set-point following in a straightforward way using only one on-line tuning parameter.

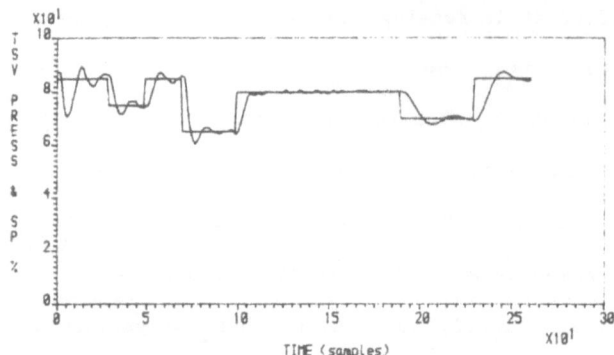

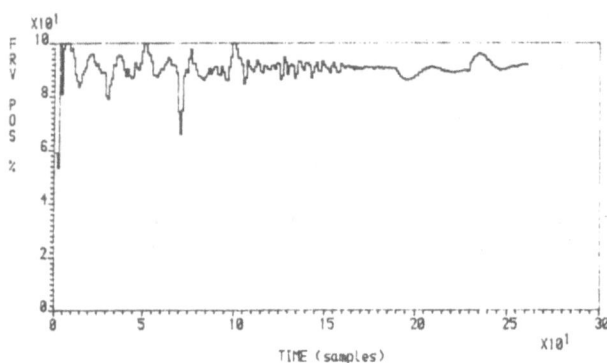

Figure 5.8 : LQG Run 1 - Overall

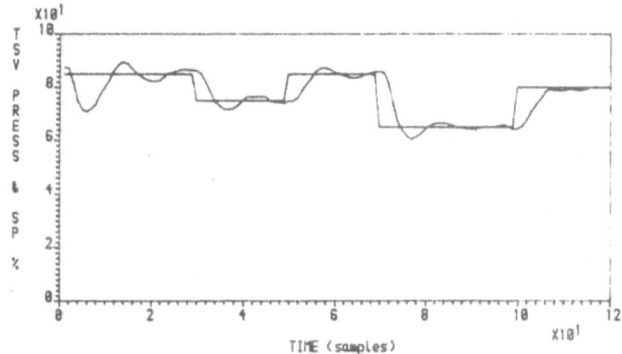

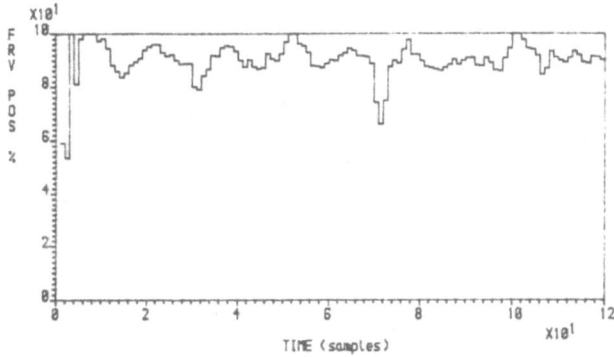

Figure 5.9 : LQG Run 1 – Tuning Phase

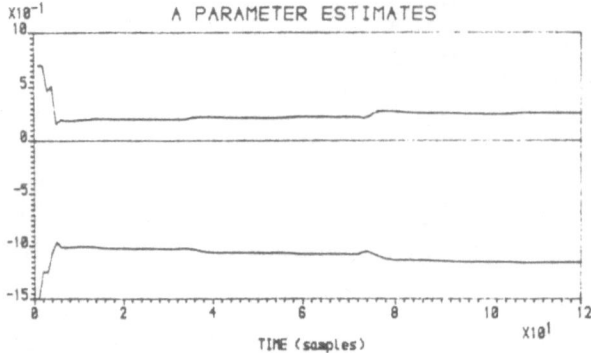

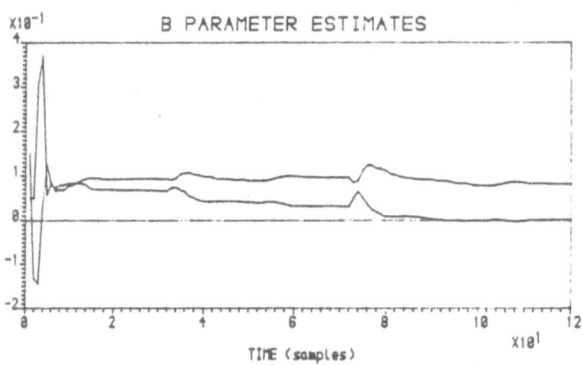

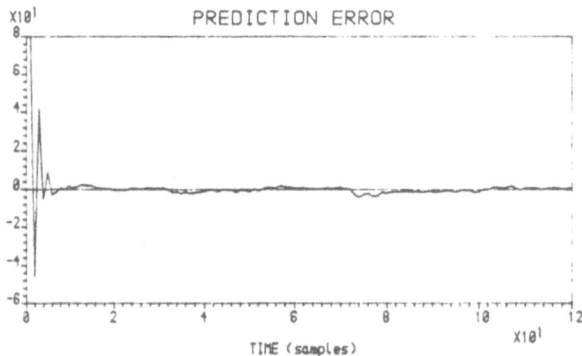

Figure 5.10 : Parameter Estimation

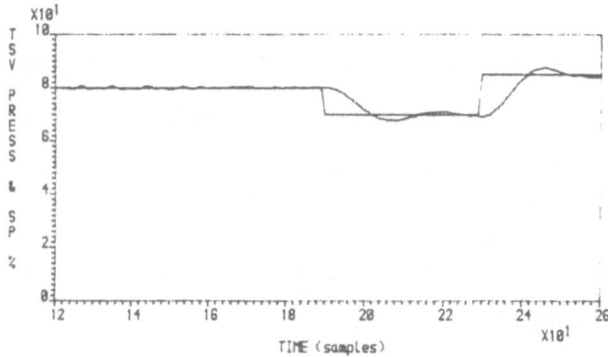

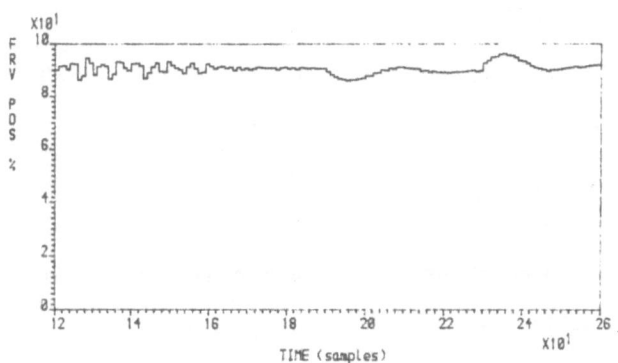

Figure 5.11 : LQG Run 1 - Fixed Phase

228

(ii)   Run 2

The second test had a duration of 315 samples (1 3/4 hours). The purpose of this test was to subject the loop to a series of step changes in set-point for different values of control weighting while monitoring the responses of the other controlled variables in the system (T1, T2 and FRVDP).

The TSV steam pressure and set-point, and the FRV position for the whole test are shown in Figure 5.12. The value of control weighting was changed after each of the down/up changes in set-point as follows:

t = 0   , μ = 0.8

t = 110 , μ = 20.0

t = 168 , μ = 50.0

In the steady-state phase following the set-point changes (t > 200) the control weighting was changed as follows:

t = 243 , μ = 20.0

t = 270 , μ = 0.5

The system responses for each downward and upward step are plotted as follows:

Figure 5.13 : downstep at t = 50   with μ = 0.8

Figure 5.14 : upstep at   t = 80   with μ = 0.8

Figure 5.15 : downstep at t = 116 with μ = 20.0

Figure 5.16 : upstep at   t = 140 with μ = 20.0

Figure 5.17 : downstep at t = 170 with μ = 50.0

Figure 5.18 : upstep at   t = 180 with μ = 50.0

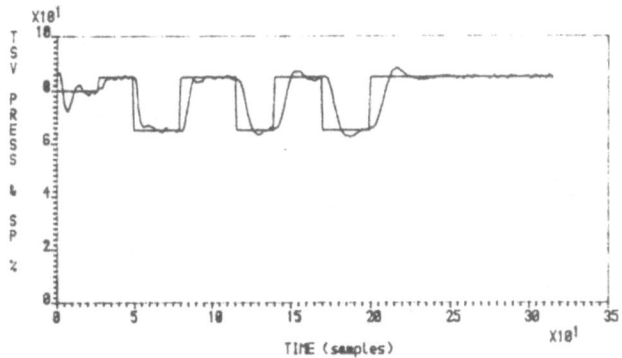

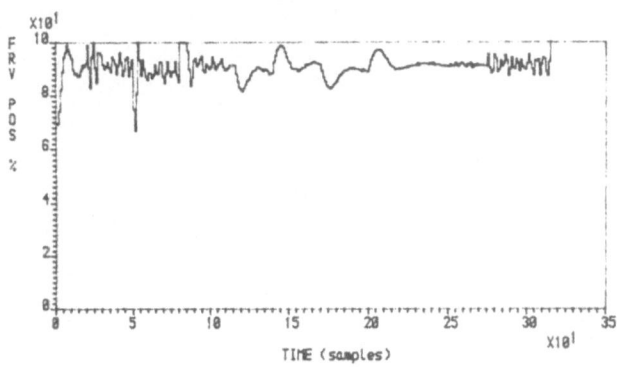

Figure 5.12 : LQG Run 2 - Overall

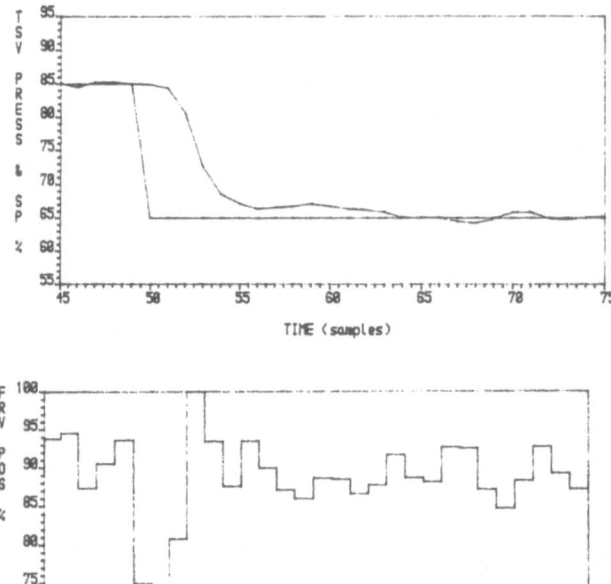

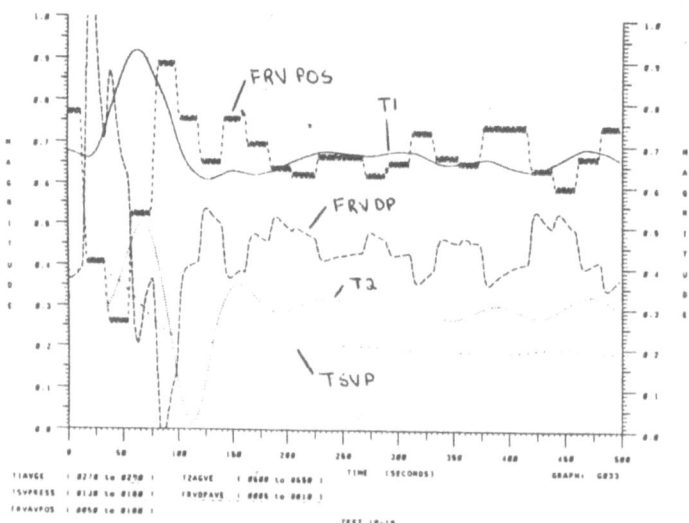

Figure 5.13 : μ = 0.8

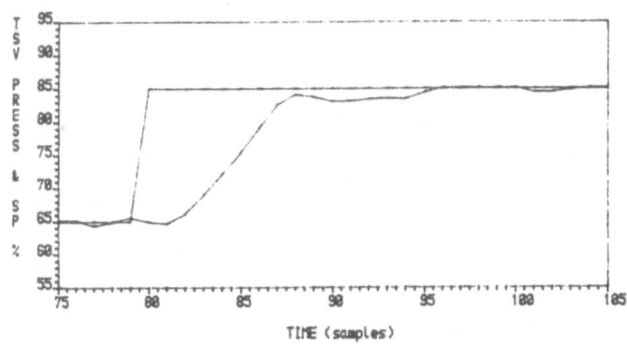

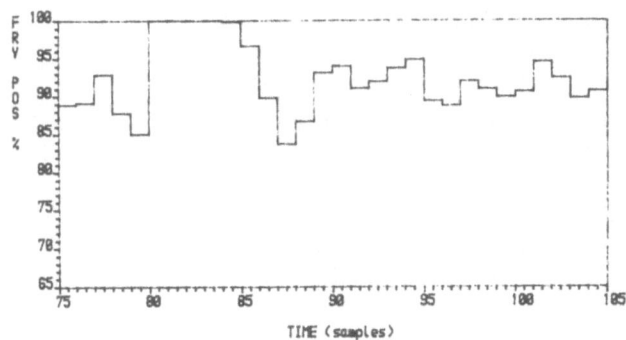

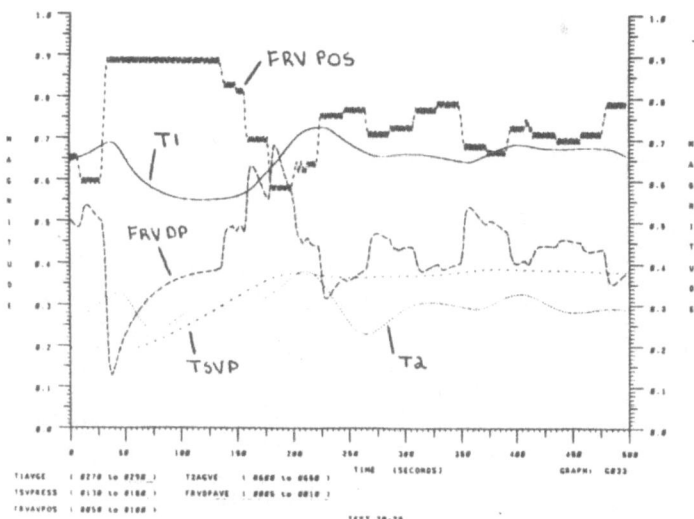

Figure 5.14 : μ = 0.8

232

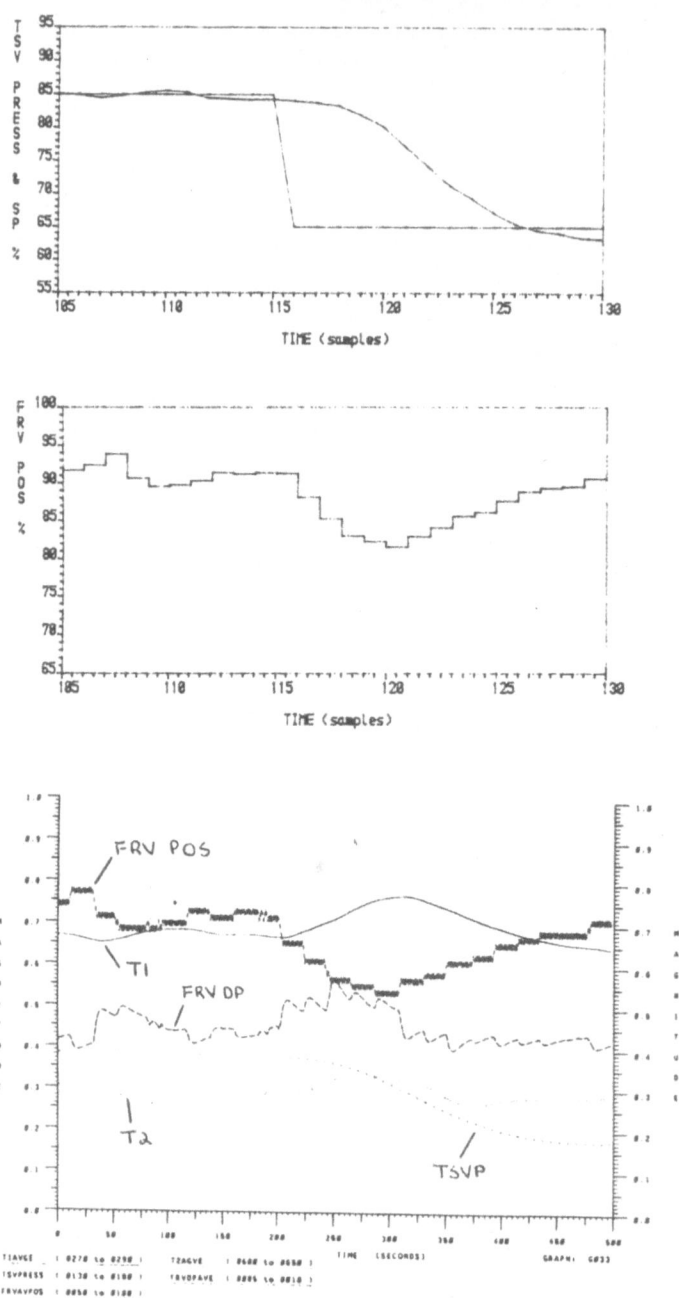

Figure 5.15 : μ = 20.0

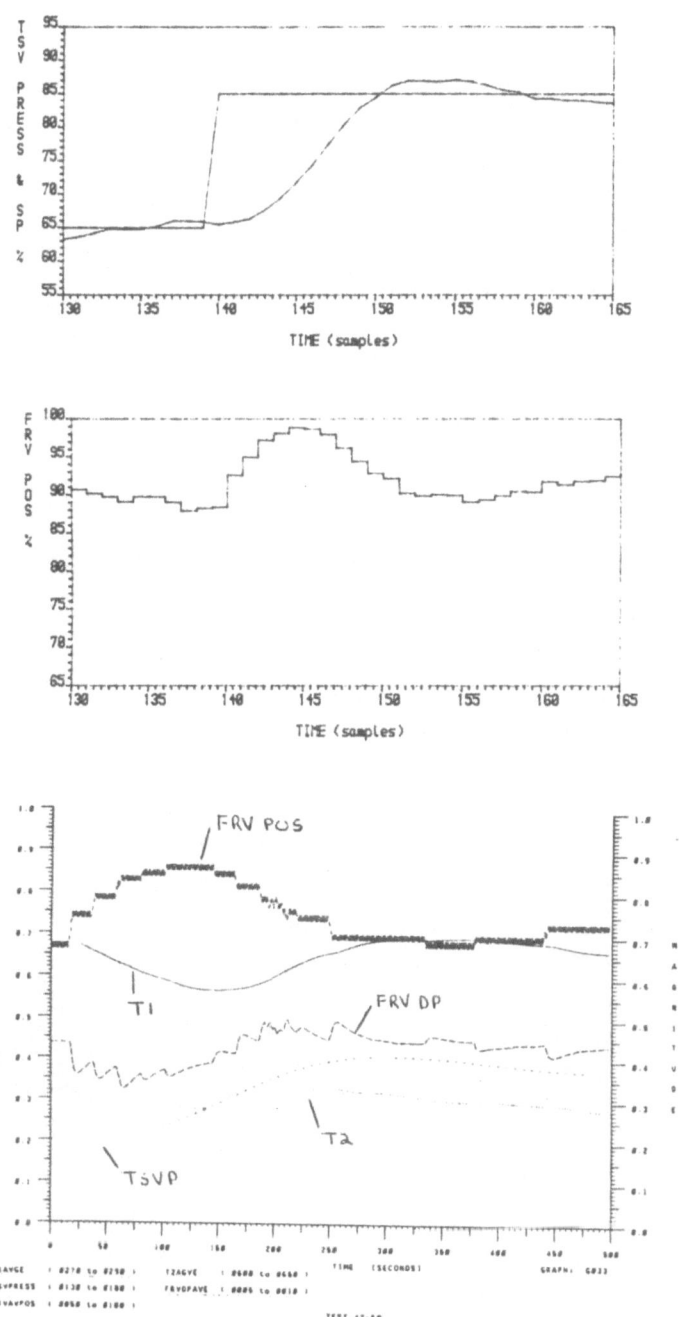

Figure 5.16 : μ = 20.0

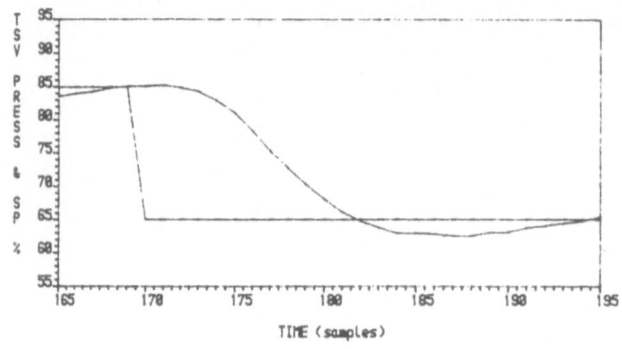

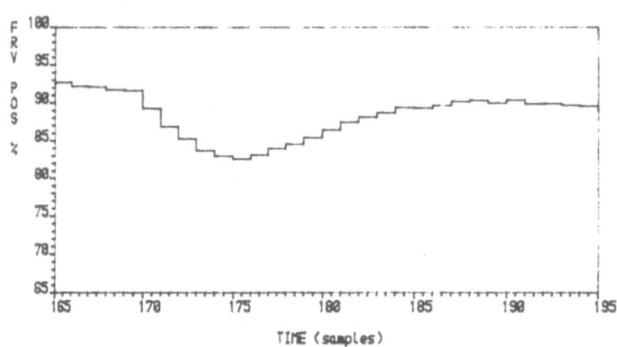

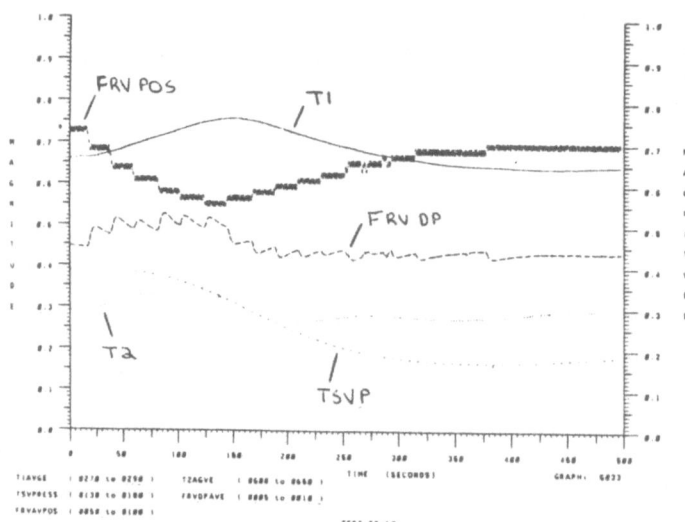

Figure 5.17 : μ = 50.0

235

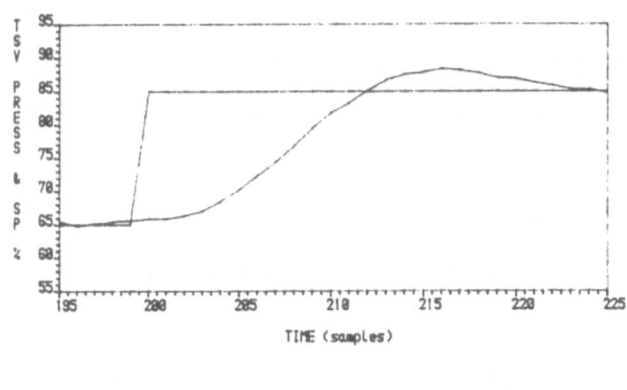

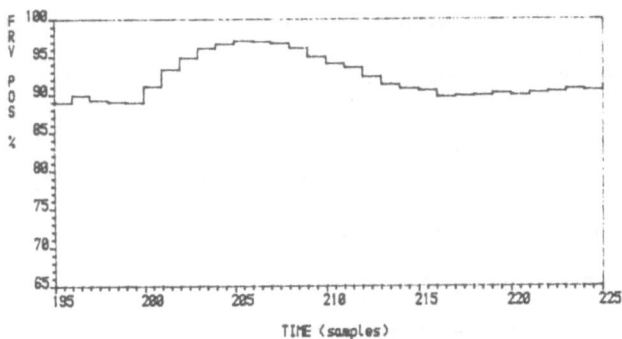

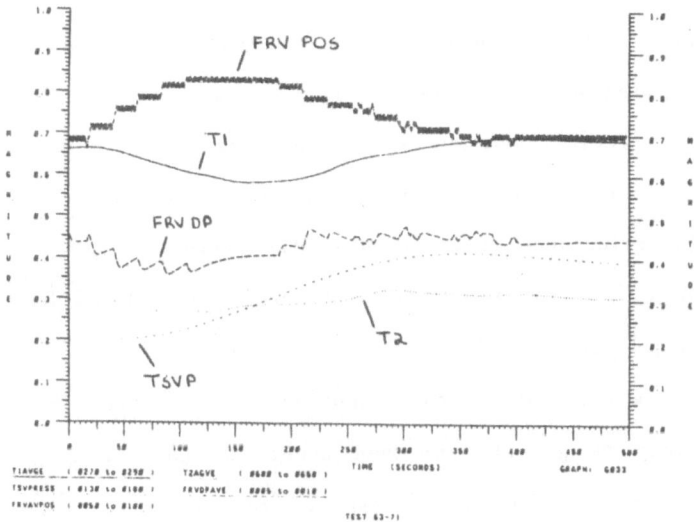

<u>Figure 5.18</u> : μ = 50.0

The responses for the steady-state period 225 < t < 255 during which the control weighting was changed from 50 to 20 (t = 243) are shown in Figure 5.19. The responses for the period 255 < t < 285 during which the control weighting was changed from 20 to 0.5 (t = 270) are shown in Figure 5.20.

From the results the following observations may be made regarding the performance of the LQG controller:

(i)   When the control weighting is <u>low</u> the loop has a very fast step response with no overshoot. However, this tight command following performance is at the cost of a relatively active actuator movement. This actuator movement also causes a severe interaction effect with the other controlled variables in the system. This interaction is perhaps only slightly worse than that observed during the PI test in Section 5.4.1.

(ii)   When the control weighting is <u>high</u> the step-response is slower with a significant overshoot. This is accompanied by a very smooth actuator movement. This smooth actuator movement leads to a dramatic reduction in the transients induced in the other control loops.

When the LQG performance is compared to the performance of the existing analogue PI controller shown in Figures 5.6 and 5.7 the following observations can be made:

(i)   For a low value of control weighting the step-response of the loop is greatly improved under LQG control : the rise-time is short and there is no overshoot. The actuator movements when μ is low are greater under LQG

237

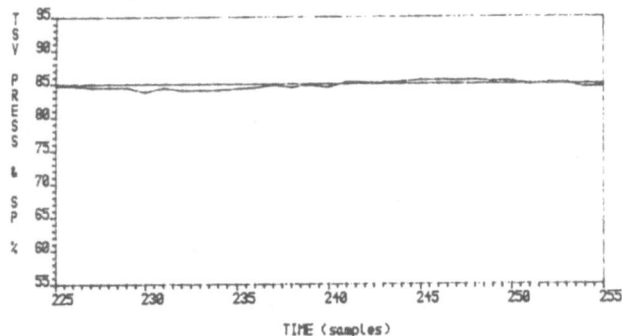

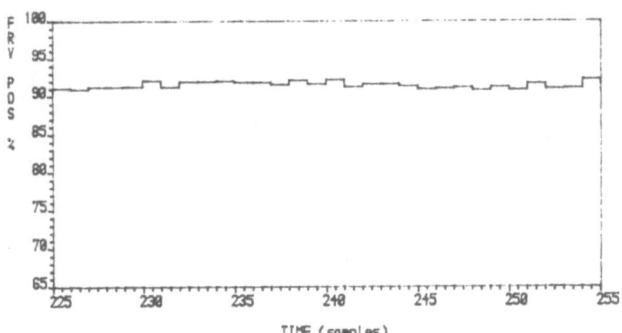

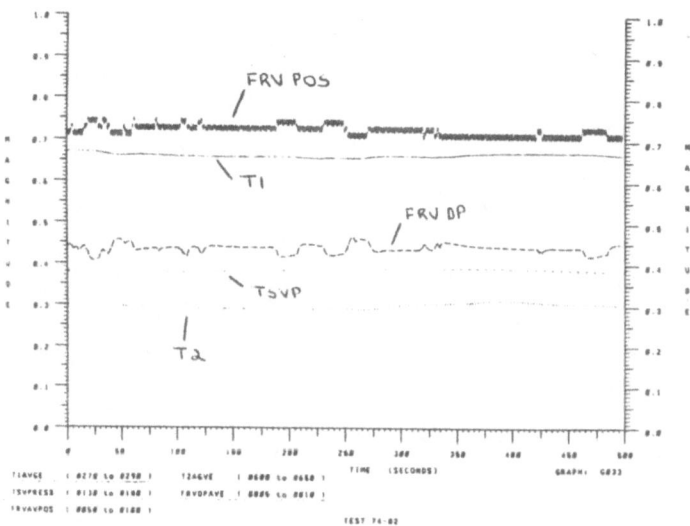

Figure 5.19 : μ = 50 → 20 (t = 243)

238

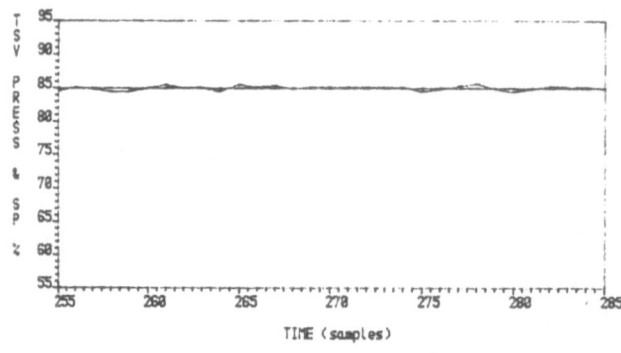

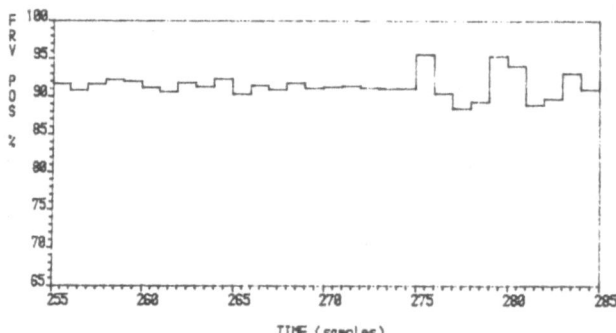

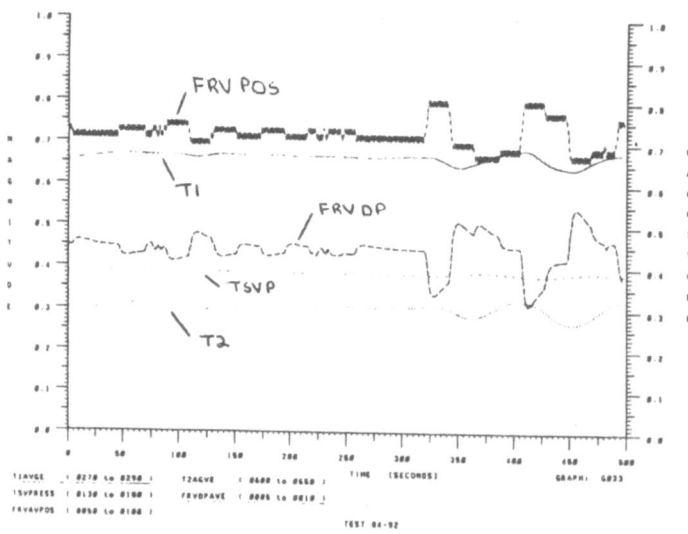

Figure 5.20 : μ = 20 → 0.5 (t = 270)

control. This may or may not be a bad thing depending on whether it is the magnitude of actuator movements or the number of actuator start/stops which is important (under LQG control the actuator only moves once every 20 seconds).

When μ is low the interaction with other loops is slightly worse under LQG control due to the magnitude of actuator movements.

(ii)   For a high value of control weighting the step-responses are similar for LQG and PI in terms of rise-time and overshoot. Under LQG control, however, the interaction transients in the other loops are dramatically reduced. This is due to the very smooth actuator movement under LQG with a large μ.

5.6   CONCLUSIONS

Although it is not usual during normal plant operation to move the TSV steam pressure set-point very frequently the tests performed still give an accurate illustration of the general interaction between loops due to various perturbation and disturbance effects. Probably the most important design factor in this particular system is to reduce these interaction effects to a minimum while still retaining an adequate level of control accuracy in each loop. In this respect the LQG controller, with a high value of control weighting, can be adjudged to give a very significant improvement in control performance when compared with the existing analogue PI controller.

The changes in the command response of the loop as the control weight μ is varied from 0.8 → 50 may be explained by plotting the locus of the closed-loop poles with respect to μ. From equation (4.37) the closed-loop poles of the transfer-function between r(t) and y(t) are determined by the zeros of the spectral factor D_c. To determine the root-locus the estimated plant was used, as follows:

$$\frac{B}{A} = \frac{d^2(0.001 + 0.08d)}{1-1.158d + 0.2548d^2}$$

The closed-loop root-locus with respect to μ is plotted in Figure 5.21. Note that the open-loop poles of the plant are 0.295 and 0.863. From Figure 5.21 it is seen that for very high μ the closed-loop poles migrate towards the open-loop poles. In the range 0.8 < μ < 50 the complex-conjugate pair of poles are seen to move significantly to the right and towards the real axis which corresponds, respectively, to reduced natural frequency of oscillation and increased damping. This corroborates the responses observed during the experiment as μ was increased.

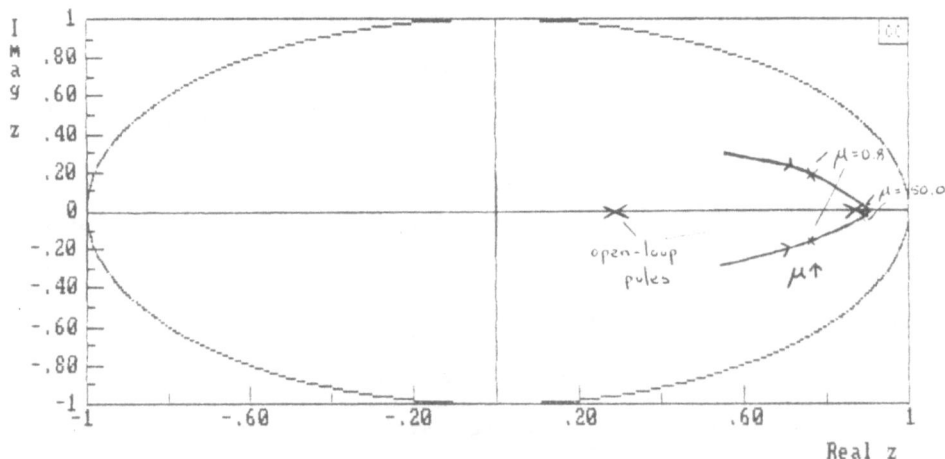

Figure 5.21 : Closed-loop Root-locus

PART FOUR

CONCLUSIONS

CHAPTER SIX

CONCLUSIONS

6.1 Stochastic optimal control theory

In application of the optimal control theory presented in
Chapter 2 the design procedure consists of choosing the cost-
function weights to achieve satisfactory performance from the
closed-loop system. In practice this is necessarily an iterative
procedure : given a plant model and a performance specification the
cost-function weights are selected, the control law is calculated and
the resulting closed-loop system is analysed with respect to
transient response, steady-state error, frequency response, stability
margins, and so on. If the performance is found not to satisfy the
specification then the cost-function weights must be altered and the
design/analysis procedure repeated. This process continues until a
satisfactory closed-loop performance is achieved.

One major area in this respect having considerable scope for
further work is the selection of the dynamic cost-function weights.
It was shown in Chapter 2 that the dynamic weights have a direct
influence on the frequency-response properties of the closed-loop
system. It was also demonstrated that fundamental design
requirements such as low gain at high frequency could be easily
introduced by this means. What is needed, however, is a
comprehensive and systematic design procedure for selection of the
dynamic weights to meet a range of performance specifications.

A criticism of the stochastic optimal control theory is that there is no direct way in which to incorporate desirable sensitivity and robustness properties into the optimisation procedure. As mentioned above, these considerations can only be partially addressed using the iterative cost-function selection procedure. Grimble (1983) and Youla and Bongiorno (1985) have augmented the standard LQG problem in an attempt to address the robustness question by incorporating sensitivity terms in the quadratic cost-function.

The H_∞ optimisation technique (Zames, 1981), on the other hand, concentrates purely on robustness properties by posing a cost-function which directly includes various system sensitivity measures. Kwakernaak (1986) and Grimble (1986d) have tackled the H_∞ design problem using polynomial optimisation techniques. The key area of research in these methods at the present time is the development of efficient and robust numerical procedures for execution of the H_∞ controller design which involves non-linear diophantine equations. If these issues can be satisfactorily resolved then the H_∞ method will become a powerful design tool for linear systems.

6.2 LQG self-tuning control

In cases where the iterative design procedure outlined above cannot be used (due to ignorance of the plant parameters, for example) a natural way to use the stochastic optimal control theory is in self-tuning control. This idea is pursued in Chapter 4 where, for identification purposes, the plant model considered is simpler than the general model of Chapter 2.

Clarke et al (1987) assert that a self-tuning control algorithm must be applicable in the following situations in order to be considered as a 'general purpose' algorithm for the stable control of the majority of real processes:

(i)   non-minimum phase plant

(ii)   open-loop unstable plant or plant with badly damped poles

(iii) a plant with variable or unknown dead-time

(iv)   a plant with unknown order

The LQG self-tuner presented in Chapter 4 overcomes all these problems.

The LQG controller has a guarantee of closed-loop stability regardless of the plant pole/zero locations. Thus, conditions (i) and (ii) above are satisfied. Since the LQG algorithm presented in Chapter 4 was based upon explicit plant identification the method can deal with variable dead-time by overparameterisation of the numerator polynomials B and D. This technique is also employed in the pole-placement self-tuners. In the pole-placement algorithms, however, overparameterisation of the plant numerator polynomials means that the order of the estimated denominator polynomial must be chosen with great care to avoid singularities in the solution of the diophantine equation caused by common factors in the estimated plant A and B polynomials. This problem does not arise in the LQG self-tuner of Chapter 4 since the couple of polynomial equations (4.22) - (4.23) can be solved (with only a small increase in computation relative to the implied equation (4.30)) to obtain the unique optimal controller regardless of any possible stable common factors in the plant A and B polynomials.

The LQG self-tuner can therefore cope with variable or unknown dead-time without suffering from overparameterisation problems and thereby satisfies conditions (iii) and (iv) above.

A further attractive feature of the LQG self-tuner is the relatively low number of on-line tuning parameters. The simple formulation of the cost-function weights given in Chapter 4 leads to the necessary inclusion in the controller of integral action, and requires the selection of only two scalar parameters (in the experimental trials presented in Chapter 5 only one parameter was actually tuned on-line).

The polynomial solution of the H_∞ control problem is another area which may be of considerable use in self-tuning systems. At the present time only a simplified version of the theory has been proposed for use in self-tuning control (Brown et al 1986, Grimble 1986e). However, developments in algorithmic aspects should give the method considerable potential as the basis for a robust self-tuning controller.

REFERENCES

Anderson, B.D.O., J.A. Gibson, and H.R. Sirisena (1985)

    Phase lag and lead weightings in linear-quadratic control

    Optimal Control Applications and Methods, 6, 249-263.

Anderson, B.D.O. and J.B. Moore (1971)

    Linear Optimal Control

    Prentice Hall, Englewood Cliffs.

Anderson, B.D.O., J.B. Moore and D.L. Mingori (1983)

    Relations between frequency dependent control and state

    weightings in LQG problems

    Proc. IEEE Conference on Decision and Control, San Antonio.

Andersson, P. (1985)

    Adaptive forgetting in recursive identification through multiple

    models

    International Journal of Control, 42, 1175.

Aseltine, J.A., A.R. Mancini and C.W. Sarture (1958)

    A survey of adaptive control systems

    IRE Trans. on Automatic Control, PGAC, Vol. 3, No. 6.

Åström, K.J. (1970)

    Introduction to Stochastic Control Theory

    Academic Press, New York.

Åström, K.J. (1980)

    Self-tuning regulator-design principles and applications

    In K.S. Narendra and R.V. Monopoli (eds),

    Applications of Adaptive Control,

    Academic Press, New York.

Åström, K.J. (1983a)

    Theory and applications of adaptive control - a survey

    Automatica, 19, 471-486.

Åström, K.J. (1983b)

    LQG self-tuners

    Proc. IFAC workshop on Adaptive Control, San Francisco.

Åström, K.J. (1983c)

    Analysis of Rohr's counter examples to adaptive control

    Proc. IEEE Conference on Decision and Control, San Antonio.

Åström, K.J. and B. Wittenmark (1973)

    On self-tuning regulators

    Automatica, 9, 185-199.

Åström, K.J. and B. Wittenmark (1974)

    Analysis of a self-tuning regulator for non-minimum phase
    systems

    Proc. IFAC Symposium on Stochastic Control, Budapest.

Åström, K.J. and B. Wittenmark (1984)

    Computer Controlled Systems

    Prentice Hall, Englewood Cliffs.

Åström, K.J. and B. Wittenmark (1985)

    The self-tuning regulators revisited

    Proc. IFAC Symposium on Identification and System Parameter
    Estimation, York.

Athans, M. and P.L. Falb (1966)

    Optimal Control

    McGraw-Hill, New York.

Bellman, R. (1957)

Dynamic Programming

Princeton University Press.

Bengtsson, G. (1973)

A theory for control of linear multivariable systems

Ph.D. Thesis, Lund, Sweden.

Bengtsson, G. (1977)

Output regulation and internal models - a frequency domain

approach

Automatica, 13, 333-345.

Bennett, S. (1979)

A History of Control Engineering 1800-1930

Peter Peregrinus, Stevenage.

Biermann, G.J. (1977)

Factorisation Methods for Discrete Sequential Estimation

Academic Press, New York.

Bode, H.W. (1940)

Relations between attenuation and phase in feedback amplifier

design

Bell system technical journal, 19, 421-454.

Brown, G., D. Biss, K.J. Hunt and M.J. Grimble (1986)

H_∞ controllers for self-tuning systems

IMA Symposium, Oxford.

Bryson, A.E. and Y.C. Ho (1969)

Applied Optimal Control

Blaisdell, Waltham, Mass.

Chang, S.S.L. (1961)

Synthesis of Optimum Control Systems

McGraw-Hill, New York.

Chen, H.F. and P.E. Caines (1984)

Adaptive linear quadratic control for discrete time systems

Proc. IFAC World Congress, Budapest.

Chen, M.J. and J.P. Norton (1987)

Estimation techniques for tracking rapid parameter changes

International Journal of Control, 45, 1387-1398.

Chestnut, H. and R. Meyer (1951-55)

Servomechanisms and Regulating System Design (Vols. 1,2)

Wiley, New York.

Clarke, D.W. and P.J. Gawthrop (1975)

Self-tuning controller

Proceedings IEE, 122, 929-934.

Clarke, D.W. and P.J. Gawthrop (1979)

Self-tuning control

Proc. IEE, 16, 633-640.

Clarke, D.W., P.P. Kanjilal and C. Mohtadi (1985)

A generalised LQG approach to self-tuning control

Int. J. Control, 41, 1509-1544.

Clarke, D.W., C. Mohtadi and P.S. Tuffs (1987)

Generalised predictive control - parts I and II

Automatica, 23, 137-160.

Cutler, C.R. and B.L. Ramaker (1980)

Dynamic matrix control - a computer control algorithm

Proc. American Control Conference, San Francisco.

De Keyser, R.M.C. and A.R. Van Cauwenberghe (1985)

Extended prediction self-adaptive control

Proc. IFAC Symposium on Identification and System Parameter

Estimation, York.

Deshpande, P.B. and R.H. Ash (1980)

Elements of Computer Process Control with Advanced Control

Applications

Instrument Society of America.

Doyle, J.C. and G. Stein (1981)

Multivariable feedback design : concepts for a classical/modern

synthesis

Trans. IEEE on Automatic Control, AC-26, 4-16.

Egardt, B. (1979)

Stability of Adaptive Controllers

Springer-Verlag, Berlin.

Egardt, B. (1979b)

Unification of some continuous-time adaptive control schemes

Trans. IEEE on Automatic Control, AC-24, 588.

Ettler, P. (1986)

An adaptive controller for Skoda twenty rolls cold rolling mills

Proc. IFAC workshop on Adaptive Systems in Control and Signal

Processing, Lund.

Fortescue, T.R., L.S. Kershenbaum and B.E. Ydstie (1981)

Implementation of self-tuning regulators with variable

forgetting factors

Automatica, 17, 831.

Francis, B.A. and W.M. Wonham (1976)

The internal model principle of control theory

Automatica, 12, 457–465.

Gadd, C.J. (1929)

Babylonian Law

Encyclopedia Britannica, (14th ed), 2, 863.

Gawthrop, P.J. (1977)

Some interpretations of the self-tuning controller

Proc. IEE, 124, 889–894.

Gawthrop, P.J. (1978)

Developments in optimal and self-tuning control theory

D. Phil thesis, Oxford, OUEL report 1239/78.

Gawthrop, P.J. (1986)

Continuous-time Self-Tuning Control

Research Studies Press, Lechworth.

Goodwin, G.C. (1985)

Some observations on robust estimation and control

Proc. IFAC Symposium on Identification and System Parameter

Estimation, York.

Goodwin, G.C., D.J. Hill and M. Palaniswami (1985)

Towards an adaptive robust controller

IFAC Symposium on Identification and System Parameter

Estimation, York.

Gregory, P.C., ed. (1959)

Proceedings of the self adaptive flight control systems

symposium

Wright Air Development Centre Technical Report 59–49, Ohio.

Grimble, M.J. (1981)

A control weighted minimum-variance controller for non-minimum

phase systems

International Journal of Control, 33, 751-762.

Grimble, M.J. (1983)

Robust LQG design of discrete systems using a dual criterion

Proc. IEEE Conference on Decision and Control, San Antonio.

Grimble, M.J. (1984)

Implicit and explicit LQG self-tuning controllers

Automatica, 20, No. 5, 661-669.

Grimble, M.J. (1986a)

Controllers for LQG self-tuning applications with coloured

measurement noise and dynamic costing

Proc. IEE, 133, Pt.D, 19-29.

Grimble, M.J. (1986b)

Feedback and feedforward LQG controller design

Proc. American Control Conference, Seattle.

Grimble, M.J. (1986c)

Convergence of explicit LQG self-tuning controllers

To appear, Proc. IEE.

Grimble, M.J. (1986d)

Optimal H_∞ robustness and the relationship to LQG design

problems

Int. J. Control, 43, 351-372.

Grimble, M.J. (1986e)

H_∞ robust controller for self-tuning control applications

ICU Report 97.

Gupta, N.A. (1980)

Frequency-shaped cost functionals = extension of linear
quadratic gaussian design methods
Journal of Guidance and Control, 3, 529-535.

Hagglund, T. (1983)

New estimation techniques for adaptive control
Ph.D thesis, Lund.

Hall, A.C. (1943)

The Analysis and Synthesis of Linear Servomechanisms
MIT Press, Cambridge, Mass.

Harris, C.J. and S.A. Billings, eds. (1981)

Self-Tuning and Adaptive Control
Peter Peregrinus, London.

Hazen, H.L. (1934)

Theory of servomechanisms
Journal of the Franklin Institute, 218, 279-331.

Hodgson, A.J.F. (1982)

Problems of integrity in applications of adaptive controllers
D.Phil Thesis, Oxford, OUEL report 1436/82.

Horowitz, I.M. (1963)

Synthesis of Feedback Systems
Academic Press, New York.

Hunt, K.J. (1988a)

A single-degree-of-freedom polynomial solution to the optimal
feedback/feedforward stochastic tracking problem
Kybernetika, 24, 81-97.

Hunt, K.J. (1988b)

General polynomial solution to the optimal feedback/feedforward
stochastic tracking problem
International Journal of Control, to appear.

Hunt, K.J. (1986)

A survey of recursive identification algorithms

Trans. Institute of Measurement and Control, 8, 273-278.

Hunt, K.J. and M.J. Grimble (1988)

LQG based self-tuning controllers

In K. Warwick (ed.), Implementation of Self-Tuning Controllers,

Peter Peregrinus, London.

Hunt, K.J., M.J. Grimble, M.J. Chen and R.W. Jones (1986)

Industrial LQG self-tuning controller design

Proc. IEEE conference on decision and control, Athens.

Hunt, K.J., M.J. Grimble and R.W. Jones (1986)

Developments in LQG self-tuning control

IEE colloquium on advances in adaptive control, London.

Hunt, K.J., M.J. Grimble and R.W. Jones (1987)

LQG feedback and feedforward self-tuning control

Proc. IFAC World Congress, Munich.

Hunt, K.J. and R.W. Jones (1988)

Personal-computer based self-tuning controller

Journal of Microcomputer Applications, 11, 95-106.

Hunt, K.J. and M. Šebek (1989)

Optimal multivariable regulation with disturbance measurement

feedforward

International Journal of Control, to appear.

Hunt, K.J., M. Šebek and M.J. Grimble (1987)

Optimal multivariable LQG control using a single diophantine

equation

International Journal of Control, 46, 1445-1453.

Hurwitz, A. (1985)

Uber die Bedingungen, unter welchen eine Gleichung nur Wurzelm
mit negativen reelen Teilen besitzt

Math. Annalen, 273-284.

Irving, E. (1979)

Improving power network stability and unit stress with adaptive
generator control

Automatica, 15, 31-46.

Jacobs, O.L.R. (1961)

A review of self-adjusting systems in automatic control

Journal of Electronics and Control, 10, 311-322.

James, H.M., N.B. Nichols and R.S. Phillips, eds. (1947)

Theory of Servomechanisms

McGraw-Hill, New York.

Ježek, J. (1982)

New algorithm for minimal solution of linear polynomial
equations

Kybernetika, 18, 505.

Ježek, J. and V. Kučera (1985)

Efficient algorithm for matrix spectral factorisation

Automatica, 21, 663.

Kalman, R.E. (1958)

Design of a self-optimising control system

Trans. ASME, 80, 440-449.

Kalman, R.E. (1960)

Contributions to the theory of optimal control

Bol. Soc. Math. Mexicana, 5, 102-119.

Kalman, R.E. (1963)

Mathematical description of linear dynamical systems

SIAM Journal on Control, 1, 152-192.

Kalman, R.E. and R.S. Bucy (1961)

New results in linear filtering and prediction theory

Trans. ASME, 83, SER.D, 95-108.

Kalman, R.E., P.L. Falb and M.A. Arbib (1969)

Topics in Mathematical Systems Theory

McGraw-Hill, New York.

Kučera, V. (1974)

Algebraic theory of discrete optimal control for multivariable

systems

Kybernetika, 10-12, 1-240 (supplement).

Kučera, V. (1975)

Stability of discrete linear feedback systems

Proc. IFAC world congress, Boston.

Kučera, V. (1979)

Discrete Linear Control

Wiley, Chichester.

Kučera, V. (1984)

The LQG control problem : a study of common factors

Problems of Control and Information Theory, 13, 239-251.

Kučera, V. (1987)

Private communication.

Kučera, V. and M. Šebek (1984a)

A polynomial solution to regulation and tracking -

Part 1 : Deterministic problem

Kybernetika, 20, 177-188.

Kučera, V. and M. Šebek (1984b)

    A polynomial solution to regulation and tracking –

    Part 2 : Stochastic problem

    Kybernetika, 20, 257–282.

Kulhavy, R. and M. Karny (1984)

    Tracking of slowly varying parameters by directional

    forgetting

    Proc. IFAC World Congress, Budapest.

Kwakernaak, H. and R. Sivan (1972)

    Linear Optimal Control Systems

    Wiley, New York.

Kwakernaak, H. (1986)

    A polynomial approach to minimax frequency domain optimization

    of multivariable feedback systems

    Int. J. Control, 44, 117–156.

Lam, K.P. (1980)

    Implicit and explicit self-tuning controllers

    D. Phil thesis, Oxford.

Landau, I.D. (1979)

    Adaptive Control – The Model Reference Approach

    Marcel Dekker, New York.

Lizr, A. (1986)

    Linear-quadratic self-tuning regulators in paper machine control

    systems

    Proc. IFAC workshop on Adaptive Systems in Control and Signal

    Processing, Lund, Sweden.

Ljung, L. and T. Söderström (1983)

    Theory and Practice of Recursive Identification

    MIT Press, Cambridge, Mass.

MacFarlane, A.G.J. (1979)

 The development of frequency-response methods in automatic
 control
 Trans. IEEE, AC-24, 250-265.

MacFarlane, A.G.J. and N. Karcanias (1976)

 Poles and zeros of multivariable systems : A survey of the
 algebraic, geometric and complex-variable theory
 International Journal of Control, 24, 33-74.

Maxwell, J.C. (1868)

 On governors
 Proc. Royal Society of London, 16, 270-283.

Mayne, D.Q. (1973)

 The design of linear multivariable systems
 Automatica, 9, 201-207.

Mayr, O. (1970)

 The Origins of Feedback Control
 MIT press, Cambridge, Mass.

Minorsky, N. (1922)

 Directional stability of automatically steered bodies
 Journal of the American Society of Naval Engineers, 42,
 280-309.

Mishkin, E. and L. Braun, eds. (1961)

 Adaptive Control Systems
 McGraw-Hill, New York.

Moore, J.B. (1984)

 A globally convergent recursive adaptive LQG regulator
 Proc. IFAC World Congress, Budapest.

Mosca, E., G. Zappa and C. Manfredi (1984)

Multistep horizon self-tuning controllers : the MUSMAR approach

Proc. IFAC World Congress, Budapest.

Narendra, K.S. and R.V. Monopoli, eds (1980)

Applications of Adaptive Control

Academic Press, New York.

Newton, G.C., L.A. Gould and J.F. Kaiser (1957)

Analytical Design of Linear Feedback Controls

Wiley, New York.

Norton, J.P. (1986)

An Introduction to Identification

Academic Press, London.

Nyquist, H. (1932)

Regeneration theory

Bell system technical journal, 11, 126-147.

Parks, P.C., W. Schaufelberger, Chr. Schmid and H. Unbehauen (1980)

Application of adaptive control systems

In H. Unbehauen (ed), Methods and Applications in Adaptive

Control

Springer-Verlag, Berlin.

Peterka , V. (1970)

Adaptive digital regulation of noisy systems

Proc. IFAC symposium on Identification and System Parameter

Estimation, Prague.

Peterka , V. (1972)

On steady-state minimum variance control strategy

Kybernetika, 8, 219-232.

Peterka , V. (1984)

Predictor based self-tuning control

Automatica, 20, 39-50.

Peterka , V. (1986)

Control of uncertain processes : Applied theory and algorithms

Kybernetika, 22, No. 3-6 (supplement in four parts).

Pontryagin, L.S., V.G. Boltyanskii, R.V. Gamkrelidze and Y.F.

Mischensko (1963)

The Mathematical Theory of Optimal Processes

Interscience, New York.

Postlethwaite, I. and A.G.J. MacFarlane (1979)

Complex Variable Approach to the Analysis of Linear

Multivariable Feedback Systems

Springer-Verlag, Berlin.

Richalet, J., A. Rault, J.L. Testud and J. Papon (1978)

Model predictive heuristic control : applications to industrial

processes

Automatica, 14, 413-428.

Roberts, A.P. (1986)

Simpler polynomial solutions in control and filtering

IMA Journal of Mathematical Control and Information, 3,

311-321.

Roberts, A.P. (1987a)

Simpler polynomial solutions in stochastic feedback control

International Journal of Control, 45, 117-126.

Roberts, A.P. (1987b)

Stochastic regulator with dynamic cost weighting

International Journal of Control, 45, 1103-1111.

Roberts, A.P. (1987c)

Stability problems with a stochastic regulator

International Journal of Control, 45, 1237-1242.

Roberts, A.P. (1987d)

Generalised polynomial optimization of stochastic feedback
control

International Journal of Control, 45, 1243-1254.

Roberts, A.P. (1987e)

Polynomial optimisation of stochastic feedback control for
unstable plants

International Journal of Control, 45, 1953-1961.

Rohrs, C.E., L. Valavani, M. Athans and G. Stein (1982)

Robustness of adaptive control algorithms in the presence of
unmodelled dynamics

Proc. IEEE Conference on Decision and Control, Orlando.

Rosenbrock, H.H. (1969)

Design of multivariable control systems using the inverse
Nyquist array

Proc. IEE, 116, 1929-1936.

Rosenbrock, H.H. (1970)

State-Space and Multivariable Theory

Nelson, London.

Routh, E.J. (1877)

A treatise on the stability of a given state of motion

Reprinted in Fuller, A.T., ed (1976). Stability of Motion,

Taylor and Francis, London.

Šebek, M. (1981)

Ph.D Thesis

Institute of Information Theory and Automation, Czechoslovak
Academy of Sciences, Prague.

Šebek, M. (1982)

Polynomial design of stochastic tracking systems

Trans. IEEE, AC-27, 468-470.

Šebek, M. (1983a)

Direct polynomial approach to discrete-time stochastic

tracking

Problems of Control and Information Theory, 12, 293-300.

Šebek, M. (1983b)

Stochastic multivariable tracking : a polynomial equation

approach

Kybernetika, 19, 453-459.

Šebek, M. (1987)

Private communication.

Šebek, M., K.J. Hunt and M.J. Grimble (1988)

LQG regulation with disturbance measurement feedforward

International Journal of Control, 47, 1497-1505.

Šebek, M. and V. Kučera (1982)

Polynomial approach to quadratic tracking in discrete linear

systems

Trans. IEEE on Automatic Control, AC-27, 1248-1250.

Shaked, U. (1976)

A general transfer-function approach to the steady-state linear

quadratic Gaussian control problem

International Journal of Control, 24, 771-800.

Shinskey, F.G. (1979)

Process Control Systems (2nd ed.)

McGraw-Hill, New York.

Sternad, M. (1985)

    Feedforward control of non-minimum-phase systems

    Report UPTEC 85104R, Uppsala University.

Sternad, M. (1987)

    Optimal and adaptive feedforward regulators

    Ph.D Thesis, Uppsala University, Sweden.

Stromer, P.R. (1959)

    Adaptive or self-optimizing control systems - a bibliography

    IRE Trans. on Automatic Control, PGAC , Vol. 4, May.

Truxal, J.G. (1955)

    Control System Synthesis

    McGraw-Hill, New York.

Truxal, J.G. (1964)

    Theory of self-adjusting systems

    Proc. IFAC World Congress, (1963), Basle, 386-392.

Tuffs, P.S. (1984)

    Self-tuning control : algorithms and applications

    D.Phil Thesis, Oxford.

Unbehauen, H., ed. (1980)

    Methods and Applications in Adaptive Control

    Springer-Verlag, Berlin.

Usher, A.P. (1954)

    A History of Mechanical Inventions (2nd ed.)

    Harvard University Press.

Van Amerongen, J. (1981)

    A model reference adaptive autopilot for ships - practical results

    Proc. IFAC World Congress, Kyoto.

Warwick, K., ed. (1988)

Implementation of Self-Tuning Controllers

Peter Peregrinus, London.

Wellstead P.E., D. Prager and P. Zanker (1979)

Pole assignment self-tuning regulator

Proc. IEE, 126, 781-787.

Wellstead, P.E. and S.P. Sanoff (1981)

Extended self-tuning algorithm

International Journal of Control, 34, 433-455.

Whitaker, H.P., J. Yamron and A. Kezer (1958)

Design of model reference adaptive control systems for aircraft

Report R-164, Instrumentation Laboratory, MIT.

Wiener, N. (1949)

Extrapolation, Interpolation and Smoothing of Stationary Time
Series

MIT Press, Cambridge, Mass.

Wolf, A. (1938)

A History of Science, Technology and Philosophy in the
Eighteenth Century

George Allen and Unwin, London.

Wolovich, W.A. (1974)

Linear Multivariable Systems

Springer, New York.

Wonham, W.M. (1974)

Linear Multivariable Control : A Geometric Approach

Springer-Verlag, Berlin.

Ydstie, B.E. (1984)

    Extended horizon adaptive control

    Proc. IFAC World Congress, Budapest.

Youla, D.C., J.J. Bongiorno and H.A. Jabr (1976a)

    Modern Wiener-Hopf design of optimal controllers – Part 1 :

    The single-input-single-output case

    Trans. IEEE on Automatic Control, AC-21, 3-13.

Youla, D.C., H.A. Jabr and J.J. Bongiorno (1976b)

    Modern Wiener-Hopf design of optimal controllers – Part 2 : The

    multivariable case

    Trans. IEEE on Automatic Control, AC-21, 319-338.

Youla, D.C. and J.J. Bongiorno (1985)

    A feedback theory of two-degree-of-freedom Wiener-Hopf design

    Trans. IEEE on Automatic Control, AC-30, 652-665.

Zadeh, L.A. and C.A. Desoer (1963)

    Linear Systems Theory : The State-Space Approach

    McGraw-Hill, New York.

Zhao-Ying, Z. and K.J. Åström (1981)

    A microprocessor implementation of an LQG self-tuner

    Internal report, Lund Institute of Technology, Sweden.

Zames, G. (1981)

    Feedback and optimal sensitivity : model reference

    transformations, multiplicative seminorms, and approximate

    inverses

    Trans. IEEE on Automatic Control, AC-26, 301-320.

Ziegler, J.G. and N.B. Nichols (1942)

    Optimum settings for automatic controllers

    Trans. ASME, 64, 759-768.

APPENDICES

APPENDIX 1 : Proof of Theorem 1

The <u>closed loop transfer function</u> M and the <u>sensitivity function</u> S for the SDF control structure are defined as:

$$M \triangleq \frac{C_c}{1 + W_p C_c} \quad , \quad S \triangleq \frac{1}{1 + W_p C_c} \tag{A1.1}$$

thus:

$$M = C_c S \quad , \quad S = 1 - W_p M \tag{A1.2}$$

From the SDF system structure shown in Figure 2.3 the control input and tracking error signals may be written as:

$$u = -M(d + n - m - W_x \psi_{\ell n}) - S C_{cf} f \tag{A1.3}$$

$$e = -(1 - W_p M)(d - m - W_x \psi_{\ell n}) + W_p M n$$

$$- \psi_{rn} - (W_x - W_p S C_{cf}) f \tag{A1.4}$$

where:

$$C_{cf} \triangleq C_{ff} + C_c W_x \tag{A1.5}$$

From equations (A1.3) and (A1.4) the control input and tracking error spectral densities may be written as:

$$\phi_u = M(\phi_d + \phi_n + \phi_m + W_x \sigma_{\ell n} W_x^*) M^* + S C_{cf} \phi_f C_{cf}^* S^* \tag{A1.6}$$

$$\phi_e = (1 - W_p M)(\phi_d + \phi_m + W_x \sigma_{\ell n} W_x^*)(1 - W_p M)^* + W_p M \phi_n M^* W_p^*$$

$$+ \sigma_{rn} + (W_x - W_p S C_{cf}) \phi_f (W_x - W_p S C_{cf})^* \tag{A1.7}$$

Denoting the integrand of the cost-function (2.43) as I, the integrand may be written:

$$I = Q_c \phi_e + R_c \phi_u \tag{A1.8}$$

Substituting the expressions for ϕ_u and ϕ_e given in equations (A1.6) and (A1.7) into equation (A1.8) the cost-function integrand may be written, after some algebraic manipulation, as:

$$I = (W_p Q_c W_p^* + R_c) S S^* (C_{cf} \phi_f C_{cf}^*$$

$$+ C_c (\phi_d + \phi_n + \phi_m + W_x \sigma_{\ell n} W_x^*) C_c^*)$$

$$+ Q_c(W_x\phi_f W_x^* + \phi_d + \phi_m + W_x\sigma_{\ell n}W_x^* + \sigma_{rn})$$

$$- Q_c\phi_f(W_x C_{cf}^* S^* W_p^* + W_p S C_{cf}W_x^*)$$

$$- Q_c(\phi_d + \phi_m + W_x\sigma_{\ell n}W_x^*)(M^* W_p^* + W_p M) \tag{A1.9}$$

To further simplify the cost expression the <u>control</u> and <u>filter</u> <u>spectral factors</u> (Y_c and Y_f, respectively) are defined by:

$$Y_c Y_c^* \triangleq W_p Q_c W_p^* + R_c \tag{A1.10}$$

$$Y_f Y_f^* \triangleq \phi_d + \phi_n + \phi_m + W_x\sigma_{\ell n}W_x^* \tag{A1.11}$$

Similarly, the measurable disturbance spectral factor Y_{fd} is defined by:

$$Y_{fd}Y_{fd}^* \triangleq \phi_f \tag{A1.12}$$

The following auxiliary spectra are defined by:

$$\phi_o \triangleq Q_c(W_x\phi_f W_x^* + \phi_d + \phi_m + W_x\sigma_{\ell n}W_x^* + \sigma_{rn}) \tag{A1.13}$$

$$\phi_{h1} \triangleq Q_c\phi_f W_p^* W_x \tag{A1.14}$$

$$\phi_{h2} \triangleq Q_c(\phi_d + \phi_m + W_x\sigma_{\ell n}W_x^*)W_p^* \tag{A1.15}$$

Substituting from equations (A1.10) – (A1.15) into equation (A1.9), the cost-function integrand may be written as:

$$I = Y_c Y_c^* S S^* (C_{cf}Y_{fd}Y_{fd}^* C_{cf}^* + C_c Y_f Y_f^* C_c^*)$$

$$+ \phi_o - \phi_{h1}C_{cf}^* S^* - \phi_{h1}^* S C_{cf} - \phi_{h2}M^* - \phi_{h2}^* M \tag{A1.16}$$

The integrand may now be split into terms which depend on each part of the controller, and terms which do not depend on the controller at all. Completing the squares in equation (A1.16) the integrand may be expressed as:

$$I = (Y_c S C_{cf}Y_{fd} - \frac{\phi_{h1}}{Y_c^* Y_{fd}^*})\ (Y_c S C_{cf}Y_{fd} - \frac{\phi_{h1}}{Y_c^* Y_{fd}^*})^*$$

$$+ (Y_c S C_c Y_f - \frac{\phi_{h2}}{Y_c^* Y_f^*})\ (Y_c S C_c Y_f - \frac{\phi_{h2}}{Y_c^* Y_f^*})^*$$

$$+ \phi_{o1} \tag{A1.17}$$

where:

$$\phi_{o1} = \phi_o - \frac{1}{Y_c Y_c^*} \left(\frac{\phi_{h1} \phi_{h1}^*}{Y_{fd} Y_{fd}^*} + \frac{\phi_{h2} \phi_{h2}^*}{Y_f Y_f^*} \right) \qquad (A1.18)$$

The term ϕ_{o1} in equation (A1.17) does not depend on the controller and does not, therefore, enter into the following cost minimisation procedure. The first two terms in equation (A1.17) depend, respectively, on the feedforward and cascade parts of the controller.

Before proceeding it is necessary to express the spectral factors of equations (A1.10)-(A1.12) in polynomial form as follows:

$$Y_c Y_c^* \triangleq \frac{D_c D_c^*}{A_c A_c^*} \qquad (A1.19)$$

$$Y_f Y_f^* \triangleq \frac{D_f D_f^*}{A_f A_f^*} \qquad (A1.20)$$

$$Y_{fd} Y_{fd}^* \triangleq \frac{D_{fd} D_{fd}^*}{A_{fd} A_{fd}^*} \qquad (A1.21)$$

Using the polynomial equation form of the system model given in Table 2.1, and using the polynomial equation form of the cost-function weights given by equation (2.45), the spectral factors may be written as:

$$Y_c Y_c^* = \frac{B_p A_r B_q B_q^* A_r^* B_p^* + A_p A_q B_r B_r^* A_q^* A_p^*}{A_p A_q A_r A_r^* A_q^* A_p^*} \qquad (A1.22)$$

$$\begin{aligned}
Y_f Y_f^* = (& A_n A_e A_x C_d \sigma_d C_d^* A_x^* A_e^* A_n^* + A_d A_e A_x C_n \sigma_n C_n^* A_x^* A_e^* A_d^* \\
& + A_d A_n A_x E_r \sigma_r E_r^* A_x^* A_n^* A_d^* + A_d A_n A_e A_x \sigma_{rn} A_x^* A_e^* A_n^* A_d^* \\
& + A_d A_n A_e C_x \sigma_{\ell n} C_x^* A_e^* A_n^* A_d^*) / (A_d A_n A_x A_e A_e^* A_x^* A_n^* A_d^*) \qquad (A1.23)
\end{aligned}$$

$$Y_{fd} Y_{fd}^* = \frac{A_\ell \sigma_{\ell n} A_\ell^* + E_\ell \sigma_\ell E_\ell^*}{A_\ell A_\ell^*} \qquad (A1.24)$$

Comparison of equations (A1.19) – (A1.21) with equations (A1.22) – (A1.24) then yields:

$$D_c D_c^* = B_p A_r B_q B_q^* A_r^* B_p^* + A_p A_q B_r B_r^* A_q^* A_p^* \tag{A1.25}$$

$$D_f D_f^* = (A_n A'_{ex} C_d \sigma_d C_d^* A'^*_{ex} A_n^* + A_{dex} C_n \sigma_n C_n^* A_{dex}^*$$
$$+ A_n A'_{xd} E_r \sigma_r E_r^* A'^*_{xd} A_n^* + A_n A_{dex} \sigma_{rn} A_{dex}^* A_n^*$$
$$+ A_n A'_{ed} C_x \sigma_{\ell n} C_x^* A'^*_{ed} A_n^*) A'_{pf} A'^*_{pf} \tag{A1.26}$$

$$D_{fd} D_{fd}^* = A_\ell \sigma_{\ell n} A_\ell^* + E_\ell \sigma_\ell E_\ell^* \tag{A1.27}$$

and:

$$A_c = A_p A_q A_r \tag{A1.28}$$

$$A_f = A_{dex} A_n A'_{pf} \tag{A1.29}$$

$$A_{fd} = A_\ell \tag{A1.30}$$

Each of the controller dependent terms in equation (A1.17) may now be simplified separately:

(1) $\underline{C_c}$ dependent term

From the plant model equations and spectral factor definitions obtain:

$$\frac{\phi_{h2}}{Y_c^* Y_f^*} = \frac{B_p^* A_r^* B_q^* B_q (D_f D_f^* - C_n \sigma_n C_n^* A'_{dexp} A'^*_{dexp} A_p A_p^*)}{A_p A_q A'_{dexp} A_n D_c^* D_f^*} \tag{A1.31}$$

The diophantine equation (2.50) allows the strictly unstable part of equation (A1.31) to be separated as follows:

$$\frac{\phi_{h2}}{Y_c^* Y_f^*} = \frac{G}{A_p A_q A'_{dexp} A_n} + \frac{Fz^{g1}}{D_c^* D_f^*} \tag{A1.32}$$

From the system equations and spectral factor definitions obtain:

$$Y_c SC_c Y_f = \frac{D_c D_f C_{cn}}{A_p A_q A'_{dexp} A_n A_r (A_p C_{cd} + B_p C_{cn})} \tag{A1.33}$$

From equations (A1.32) and (A1.33) obtain:

$$Y_c SC_c Y_f - \frac{\phi_{h2}}{Y_c^* Y_f^*} = \frac{D_c D_f C_{cn} - GA_r(A_p C_{cd} + B_p C_{cn})}{A_p A_q A'_{dexp} A_n A_r(A_p C_{cd} + B_p C_{cn})} - \frac{Fz^{g1}}{D_c^* D_f^*} \quad (A1.34)$$

Substituting from the implied cascade diophantine equation (2.58), equation (A1.34) may be expressed as:

$$Y_c SC_c Y_f - \frac{\phi_{h2}}{Y_c^* Y_f^*} = \frac{C_{cn}H - GA_r C_{cd}}{A_q A'_{dexp} A_n A_r(A_p C_{cd} + B_p C_{cn})} - \frac{Fz^{g1}}{D_c^* D_f^*} \quad (A1.35)$$

Finally, equation (A1.35) may be expressed as:

$$Y_c SC_c Y_f - \frac{\phi_{h2}}{Y_c^* Y_f^*} = T_1^+ + T_1^- \quad (A1.36)$$

where T_1^+ denotes the first term in equation (A1.35) and T_1^- denotes the second, strictly unstable, term.

(ii) C_{cf} dependent term

From the plant model equations and spectral factor definitions obtain:

$$\frac{\phi_{h1}}{Y_c^* Y_{fd}^*} = \frac{B_p^* A_r^* B_q^* B_q C_x D_{fd}}{A_q A_\ell A_x D_c^*} \quad (A1.37)$$

The diophantine equation (2.55) allows the strictly unstable part of equation (A1.37) to be separated as follows:

$$\frac{\phi_{h1}}{Y_c^* Y_{fd}^*} = \frac{X}{A_q A_\ell A_x} + \frac{Zz^{g2}}{D_c^*} \quad (A1.38)$$

From the system equations and spectral factor definitions obtain:

$$Y_c SC_{cf} Y_{fd} = \frac{D_c C_{cd} C_{cfn} D_{fd}}{C_{cfd} A_\ell A_q A_r(A_p C_{cd} + B_p C_{cn})} \quad (A1.39)$$

From equations (A1.38) and (A1.39) obtain:

$$Y_c SC_{cf} Y_{fd} - \frac{\phi_{h1}}{Y_c^* Y_{fd}^*} = \frac{D_c C_{cd} C_{cfn} D_{fd} A_x \ -XA_r C_{cfd} (A_p C_{cd} + B_p C_{cn})}{C_{cfd} A_\ell A_x A_q A_r (A_p C_{cd} + B_p C_{cn})}$$

$$- \frac{Zz g^2}{D_c^*} \qquad (A1.40)$$

Substituting from equation (2.59) this may now be written:

$$Y_c SC_{cf} Y_{fd} - \frac{\phi_{h1}}{Y_c^* Y_{fd}^*} = \frac{C_{cd} C_{cfn} D_{fd} A_x \ - \ XA_r C_{cfd} D_f}{C_{cfd} A_\ell A_x A_q A_r D_f} - \frac{Zz g^2}{D_c^*} \qquad (A1.41)$$

Finally, equation (A1.41) may be expressed as:

$$Y_c SC_{cf} Y_{fd} - \frac{\phi_{h1}}{Y_c^* Y_{fd}^*} = T_2^+ + T_2^- \qquad (A1.42)$$

where T_2^+ denotes the first term in equation (A1.41) and T_2^- denotes the second, strictly unstable, term.

Minimisation

Substituting from equations (A1.36) and (A1.42) into equation (A1.17) the cost-function integrand may be written:

$$I = (T_1^+ + T_1^-)(T_1^+ + T_1^-)^* + (T_2^+ + T_2^-)(T_2^+ + T_2^-)^* + \phi_{o1} \qquad (A1.43)$$

In equation (A1.43) the T_i^+ terms are stable and the T_i^- terms strictly unstable for $i = \{1,2\}$. In the expansion of equation (A1.43) the terms $T_i^- T_i^{+*}$ are therefore analytic in $|z| < 1$. In addition, the terms $T_i^- T_i^{+*}/z$ are also analytic in $|z| < 1$ since the T_i^- terms are strictly proper in z^{-1} (the optimality conditions). Thus, using the identity:

$$\oint_c T^- T^{+*} \frac{dz}{z} = - \oint_c T^+ T^{-*} \frac{dz}{z} \qquad (A1.44)$$

and invoking Cauchy's Theorem, the contour integrals of the cross terms $T_i^+ T_i^{-*}$, $T_i^- T_i^{+*}$ in equation (A1.43) are zero. The cost function therefore simplifies to:

$$J = \frac{1}{2\pi j} \oint_{|z|=1} \left[\sum_{i=1}^{2} (T_i^+ T_i^{+*} + T_i^- T_i^{-*}) + \phi_{ol} \right] \frac{dz}{z} \qquad (A1.45)$$

Since the terms T_i^- and ϕ_{ol} are independent of the controller the cost-function J is minimised by setting:

$$T_i^+ = 0, \quad i = \{1,2\} \qquad (A1.46)$$

(i)   Cascade controller

From equations (A1.35) and (A1.36), setting $T_1^+ = 0$ involves:

$$C_{cn} H - GA_r C_{cd} = 0 \qquad (A1.47)$$

or:

$$C_c = \frac{GA_r}{H} \qquad (A1.48)$$

(ii) Feedforward controller

From equations (A1.41) and (A1.42), setting $T_2^+ = 0$ involves:

$$C_{cd} C_{cfn} D_{fd} A_x - XA_r C_{cfd} D_f = 0 \qquad (A1.49)$$

or:

$$C_{cf} = \frac{XA_r D_f}{C_{cd} D_{fd} A_x} \qquad (A1.50)$$

Using the definition of C_{cf} in equation (A1.5), the feedforward controller becomes:

$$C_{ff} = \frac{XA_r D_f - C_{cn} C_x D_{fd}}{D_{fd} A_x C_{cd}} \qquad (A1.51)$$

Minimum Cost

Setting $T_i^+ = 0$, $i = \{1,2\}$ in equation (Al.45), the minimum cost is found to be:

$$J_{min} = \frac{1}{2\pi j} \oint_{|z|=1} \left[\sum_{i=1}^{2} (T_i^- T_i^{-*}) + \phi_{o1} \right] \frac{dz}{z} \qquad (A1.52)$$

Stability of T_i^+ Terms

Implicit in the above proof is the requirement that the T_i^+ terms are asymptotically stable for $i = \{1,2\}$. Stability of the T_i^+ terms may be demonstrated as follows:

(i) T_1^+ term

From equations (Al.35) and (Al.36) obtain:

$$T_1^+ = \frac{C_{cn}H - GA_r C_{cd}}{A_q A'_{dexp} A_n A_r D_f D_c} \qquad (A1.53)$$

By definition, A_q and A_r are strictly Hurwitz polynomials. By virtue of conditions (a), (b) and (c) in Theorem 1 A'_{dexp} is strictly Hurwitz. From Corollary 3 in Section 2.4 A_n divides both G and C_{cn}. By definition, D_f is Hurwitz, but in the limiting case when D_f has a zero on the unit circle this zero will also be in G and C_{cn} (by virtue of Corollary 5). From Lemma 1 D_c is strictly Hurwitz. T_1^+ is therefore asymptotically stable.

(ii) T_2^+ term

From equations (Al.41) and (Al.42) obtain:

$$T_2^+ = \frac{C_{cd}C_{cfn}D_{fd}A_x - XA_r C_{cfd}D_f}{C_{cfd}A_\ell A_x A_q A_r D_f} \qquad (A1.54)$$

Substituting from the implied feedforward diophantine equation (2.60) and using equation (A1.50), the expression for T_2^+ may, after some algebraic manipulation, be written as:

$$T_2^+ = \frac{XA'_{p\ell x}(H - C_{cd})}{A'_{\ell x}A_q D_f D_c} \qquad (A1.55)$$

By definition A_q is strictly Hurwitz. By virtue of condition (c) in Theorem 1 $A'_{\ell x}$ is strictly Hurwitz. By definition, D_f is Hurwitz but in the limiting case when D_f has a zero on the unit circle this zero will also be in $A'_{p\ell x}$ (by virtue of Corollary 5 and equation (2.27)). From Lemma 1 D_c is strictly Hurwitz. T_2^+ is therefore asymptotically stable.

Solvability conditions

It only remains to relate the conditions (a)-(d) in Theorem 1 to solvability of the optimal control problem. Problem solvability in this context is taken to mean that the given data generate a controller which renders the cost-function finite.

Clearly, the cost will be finite if and only if the twelve transfer-functions in equations (2.32) and (2.33) are asymptotically stable.

Consider the case when α has a zero on the unit circle as discussed in Corollary 5. Such a zero arises when A_p has a zero on the unit circle <u>and</u> when A_d, A_x and A_e do not. Using Corollary 5 D_f and C_{cn} will also have this zero. From equation (A1.51) C_{ffn} will also have this zero. As a consequence, this unstable zero will cancel in all the transfer-functions in equations (2.32) and (2.33) and the problem remains solvable.

By Corollary 3 (which results from condition (d)) the transfer functions $B_p C_{cn} C_n / A_n$ and $A_p C_{cn} C_n / A_n$ are asymptotically stable. From equation (A1.51), and by Corollary 4 and Lemma 1, the transfer-functions $B_p C_{ffn} C_{cd} / C_{ffd}$ and $A_p C_{ffn} C_{cd} / C_{ffd}$ are asymptotically stable. Finally, conditions (a) – (c) ensure asymptotic stability of the remaining transfer-functions as follows:

(i) Condition (a)

Clearly, condition(a) ensures asymptotic stability of the transfer functions $C_d A_p C_{cd} / A_d$ and $C_d A_p C_{cn} / A_d$.

(ii) Condition (b)

Clearly, condition (b) ensures asymptotic stability of the transfer-functions $A_p C_{cd} E_r / A_e$ and $A_p C_{cn} E_r / A_e$.

(iii) Condition (c)

The fifth transfer-function in equation (2.32) is:

$$\frac{(C_x A_p C_{ffd} - B_p C_{ffn} A_x) C_{cd} E_\ell}{A_x A_\ell C_{ffd} \alpha}$$

Substituting from the implied feedforward equation (2.60) and using equation (A1.51) this may be rewritten, after some algebraic manipulation, as:

$$\frac{A_q Y E_\ell D_f}{A'_{\ell x} D_{fd} \alpha}$$

By condition (c) $A'_{\ell x}$ is strictly Hurwitz and by Lemma 1 so is D_{fd}. This transfer-function is therefore asymptotically stable.

The fifth transfer-function in equation (2.33) is:

$$\frac{(C_{cn}C_xC_{ffd} + C_{ffn}A_xC_{cd})A_pE_{\ell}}{A_xA_{\ell}C_{ffd}\alpha}$$

Substituting from equation (A1.51) this may be written:

$$\frac{(C_{cn}C_xD_{fd} + C_{ffn})A'_{p\ell x}E_{\ell}}{A'_{\ell x}D_{fd}\alpha}$$

Again, condition (c) and Lemma 1 ensure asymptotic stability of this transfer-function.

APPENDIX 2 : <u>Proof of Theorem 6</u>

The <u>closed-loop- transfer-function</u> M and the <u>sensitivity</u>

<u>function</u> S for the 2DF control structure are defined as:

$$M \triangleq \frac{C_{fb}}{1 + C_{fb}W_p} \quad , \quad S \triangleq \frac{1}{1 + C_{fb}W_p} \qquad (A2.1)$$

thus:

$$M = C_{fb}S \quad , \quad S = 1 - W_p M \qquad (A2.2)$$

From the 2DF system structure shown in Figure 2.4 the control input

and tracking error signals may now be written as:

$$u = -Md - Mn + SC_r m - SC_{1f} f + MW_x \psi_{\ell n} \qquad (A2.3)$$

$$e = -(1 - W_p M)d + W_p Mn + (1 - SW_p C_r)m - \psi_{rn}$$

$$\quad - (W_x - SW_p C_{1f})f + (1 - W_p M)W_x \psi_{\ell n} \qquad (A2.4)$$

where:

$$C_{1f} \triangleq C_{ff} + C_{fb}W_x \qquad (A2.5)$$

From equations (A2.3) and (A2.4) the control input and tracking error

spectral densities may be written as:

$$\phi_u = M(\phi_d + \phi_n + W_x \sigma_{\ell n} W_x^*)M^* + SC_r \phi_m C_r^* S^*$$

$$\quad + SC_{1f} \phi_f C_{1f}^* S^* \qquad (A2.6)$$

$$\phi_e = (1 - W_p M)(\phi_d + W_x \sigma_{\ell n} W_x^*)(1 - W_p M)^* + W_p M \phi_n M^* W_p^*$$

$$\quad + (1 - SW_p C_r)\phi_m(1 - SW_p C_r)^* + \sigma_{rn}$$

$$\quad + (W_x - SW_p C_{1f})\phi_f(W_x - SW_p C_{1f})^* \qquad (A2.7)$$

Denoting the integrand of the cost-function (2.43) as I, the

integrand may be written:

$$I = Q_c \phi_e + R_c \phi_u \qquad (A2.8)$$

Substituting the expressions for ϕ_u and ϕ_e given in equations (A2.6) and (A2.7) into equation (A2.8) the cost-function integrand may be written, after some algebraic manipulation, as:

$$I = (W_p Q_c W_p^* + R_c) \, SS^* (C_{1f} \phi_f C_{1f}^* + C_r \phi_m C_r^*$$

$$+ C_{fb}(\phi_d + \phi_n + W_x \sigma_{\ell n} W_x^*) C_{fb}^*)$$

$$+ Q_c (W_x \phi_f W_x^* + \phi_d + W_x \sigma_{\ell n} W_x^* + \phi_m + \sigma_{rn})$$

$$- Q_c \phi_f (W_x C_{1f}^* W_p^* S^* + SW_p C_{1f} W_x^*)$$

$$- Q_c \phi_m (C_r^* W_p^* S^* + SW_p C_r)$$

$$- Q_c (\phi_d + W_x \sigma_{\ell n} W_x^*)(M^* W_p^* + W_p M) \tag{A2.9}$$

To further simplify the cost expression the <u>control</u> and <u>filter spectral factors</u> (Y_c and Y_f, respectively) are defined by:

$$Y_c Y_c^* \triangleq W_p Q_c W_p^* + R_c \tag{A2.10}$$

$$Y_f Y_f^* \triangleq \phi_d + \phi_n + W_x \sigma_{\ell n} W_x^* \tag{A2.11}$$

Similarly, the <u>measurable disturbance spectral factor</u> Y_{fd} and the <u>reference spectral factor</u> Y_m are defined by:

$$Y_{fd} Y_{fd}^* \triangleq \phi_f \tag{A2.12}$$

$$Y_m Y_m^* \triangleq \phi_m \tag{A2.13}$$

The following <u>auxiliary spectra</u> are defined by:

$$\phi_o \triangleq Q_c (W_x \phi_f W_x^* + \phi_d + W_x \sigma_{\ell n} W_x^* + \phi_m + \sigma_{rn}) \tag{A2.14}$$

$$\phi_{h1} \triangleq Q_c \phi_f W_p^* W_x \tag{A2.15}$$

$$\phi_{h2} \triangleq Q_c \phi_m W_p^* \tag{A2.16}$$

$$\phi_{h3} \triangleq Q_c (\phi_d + W_x \sigma_{\ell n} W_x^*) W_p^* \tag{A2.17}$$

Substituting from equations (A2.10)-(A2.17) into equation (A2.9), the cost-function integrand may be expressed as:

$$I = Y_c Y_c^* SS^* (C_{1f} Y_{fd} Y_{fd}^* C_{1f}^* + C_r Y_m Y_m^* C_r^* + C_{fb} Y_f Y_f^* C_{fb}^*)$$
$$+ \phi_o - \phi_{h1} C_{1f}^* S^* - \phi_{h1} SC_{1f}$$
$$- \phi_{h2} C_r^* S^* - \phi_{h2} SC_r - \phi_{h3} M^* - \phi_{h3}^* M \qquad (A2.18)$$

The integrand may now be split into terms which depend on each part
of the controller, and terms which do not depend on the controller at
all. Completing the squares in equation (A2.18) the integrand may be
expressed as:

$$I = (Y_c SC_{1f} Y_{fd} - \frac{\phi_{h1}}{Y_c^* Y_{fd}^*})(Y_c SC_{1f} Y_{fd} - \frac{\phi_{h1}}{Y_c^* Y_{fd}^*})^*$$

$$+ (Y_c SC_r Y_m - \frac{\phi_{h2}}{Y_c^* Y_m^*})(Y_c SC_r Y_m - \frac{\phi_{h2}}{Y_c^* Y_m^*})^*$$

$$+ (Y_c MY_f - \frac{\phi_{h3}}{Y_c^* Y_f^*})(Y_c MY_f - \frac{\phi_{h3}}{Y_c^* Y_f^*})^*$$

$$+ \phi_{ol} \qquad (A2.19)$$

where:

$$\phi_{ol} = \phi_o - \frac{1}{Y_c Y_c^*}(\frac{\phi_{h1} \phi_{h1}^*}{Y_{fd} Y_{fd}^*} + \frac{\phi_{h2} \phi_{h2}^*}{Y_m Y_m^*} + \frac{\phi_{h3} \phi_{h3}^*}{Y_f Y_f^*}) \qquad (A2.20)$$

The term ϕ_{ol} in equation (A2.19) does not depend on the controller
and does not, therefore, enter into the following cost minimisation
procedure. The first three terms in equation (A2.19) depend,
respectively, on the feedforward, reference and feedback parts of the
controller.

Before proceeding it is necessary to express the spectral
factors of equations (A2.10) - (A2.13) in polynomial form as
follows:

$$Y_c Y_c^* \triangleq \frac{D_c D_c^*}{A_c A_c^*} \qquad (A2.21)$$

$$Y_f Y_f^* \triangleq \frac{D_f D_f^*}{A_f A_f^*} \tag{A2.22}$$

$$Y_{fd} Y_{fd}^* \triangleq \frac{D_{fd} D_{fd}^*}{A_{fd} A_{fd}^*} \tag{A2.23}$$

$$Y_m Y_m^* \triangleq \frac{D_m D_m^*}{A_m A_m^*} \tag{A2.24}$$

Using the polynomial equation form of the system model given in Table
2.1, and using the polynomial equation form of the cost-function
weights given by equation (2.45), the spectral factors may be written
as:

$$Y_c Y_c^* = \frac{B_p A_r B_q B_q^* A_r^* B_p^* + A_p A_q B_r B_r^* A_q^* A_p^*}{A_p A_q A_r A_r^* A_q^* A_p^*} \tag{A2.25}$$

$$Y_f Y_f^* = \frac{A_n A_x C_d \sigma_d C_d^* A_x^* A_n^* + A_d A_x C_n \sigma_n C_n^* A_x^* A_d^* + A_d A_n C_x \sigma_{\ell n} C_x^* A_n^* A_d^*}{A_d A_n A_x A_x^* A_n^* A_d^*} \tag{A2.26}$$

$$Y_{fd} Y_{fd}^* = \frac{A_\ell \sigma_{\ell n} A_\ell^* + E_\ell \sigma_\ell E_\ell^*}{A_\ell A_\ell^*} \tag{A2.27}$$

$$Y_m Y_m^* = \frac{A_e \sigma_{rn} A_e^* + E_r \sigma_r E_r^*}{A_e A_e^*} \tag{A2.28}$$

Comparison of equations (A2.21)-(A2.24) with equations
(A2.25)-(A2.28) then yields:

$$D_c D_c^* = B_p A_r B_q B_q^* A_r^* B_p^* + A_p A_q B_r B_r^* A_q^* A_p^* \tag{A2.29}$$

$$D_f D_f^* = (A_n A_x' C_d \sigma_d C_d^* A_x'^* A_n^* + A_{dx} C_n \sigma_n C_n^* A_{dx}^* + A_d' A_n C_x \sigma_{\ell n} C_x^* A_n^* A_d'^*) A_p' A_p'^* \tag{A2.30}$$

$$D_{fd}D_{fd}^* = A_\ell \sigma_{\ell n} A_\ell^* + E_\ell \sigma_\ell E_\ell^* \tag{A2.31}$$

$$D_m D_m^* = A_e \sigma_{rn} A_e^* + E_r \sigma_r E_r^* \tag{A2.32}$$

and:

$$A_c = A_p A_q A_r \tag{A2.33}$$

$$A_f = A_{dx} A_n A_p' \tag{A2.34}$$

$$A_{fd} = A_\ell \tag{A2.35}$$

$$A_m = A_e \tag{A2.36}$$

Each of the controller dependent terms in equation (A2.19) may now be simplified separately:

(i) C_{fb} dependent term

From the plant model equations and spectral factor definitions obtain:

$$\frac{\phi_{h3}}{Y_c^* Y_f^*} = \frac{B_p^* A_r^* B_q^* B_q (D_f D_f^* - C_n \sigma_n C_n^* A_{dx}' A_{dx}'^* A_p A_p^*)}{A_p A_q A_{dx}' A_n D_c^* D_f^*} \tag{A2.37}$$

The diophantine equation (2.78) allows the strictly unstable part of equation (A2.37) to be separated as follows:

$$\frac{\phi_{h3}}{Y_c^* Y_f^*} = \frac{G}{A_p A_q A_{dx}' A_n} + \frac{Fz^{g1}}{D_c^* D_f^*} \tag{A2.38}$$

From the system equations and spectral factor definitions obtain:

$$Y_c M Y_f = \frac{D_c D_f C_{fbn}}{A_p A_q A_{dx}' A_n A_r (A_p C_{fbd} + B_p C_{fbn})} \tag{A2.39}$$

From equations (A2.38) and (A2.39) obtain:

$$Y_c M Y_f - \frac{\phi_{h3}}{Y_c^* Y_f^*} = \frac{D_c D_f C_{fbn} - G A_r (A_p C_{fbd} + B_p C_{fbn})}{A_p A_q A_{dx}' A_n A_r (A_p C_{fbd} + B_p C_{fbn})} - \frac{Fz^{g1}}{D_c^* D_f^*} \tag{A2.40}$$

Substituting from the implied feedback diophantine equation (2.89), equation (A2.40) may be expressed as:

$$Y_c MY_f - \frac{\phi_{h3}}{Y_c^* Y_f^*} = \frac{C_{fbn}H - C_{fbd}GA_r}{A_q A_{dx}' A_n A_r (A_p C_{fbd} + B_p C_{fbn})} - \frac{Fz^{g1}}{D_c^* D_f^*} \qquad (A2.41)$$

Finally, equation (A2.41) may be expressed as:

$$Y_c MY_f - \frac{\phi_{h3}}{Y_c^* Y_f^*} = T_1^+ + T_1^- \qquad (A2.42)$$

where T_1^+ denotes the first term in equation (A2.41) and T_1^- denotes

the second, strictly unstable, term.

(ii) $\underline{C_r \text{ dependent term}}$

From the plant model equations and spectral factor definitions

obtain:

$$\frac{\phi_{h2}}{Y_c^* Y_m^*} = \frac{B_p A_r^* B_q^* B_q D_m}{A_q A_e A_c^*} \qquad (A2.43)$$

The diophantine equation (2.83) allows the strictly unstable part of

equation (A2.43) to be separated as follows:

$$\frac{\phi_{h2}}{Y_c^* Y_m^*} = \frac{M}{A_q A_e} + \frac{Nz^{g2}}{D_c^*} \qquad (A2.44)$$

From the system equations and spectral factor definitions obtain:

$$Y_c SC_r Y_m = \frac{D_c C_{fbd} D_m C_{rn}}{C_{rd} A_e A_q A_r (A_p C_{fbd} + B_p C_{fbn})} \qquad (A2.45)$$

From equations (A2.44) and (A2.45) obtain:

$$Y_c SC_r Y_m - \frac{\phi_{h2}}{Y_c^* Y_m^*} = \frac{D_c C_{rn} D_m C_{fbd} - C_{rd} MA_r (A_p C_{fbd} + B_p C_{fbn})}{C_{rd} A_e A_q A_r (A_p C_{fbd} + B_p C_{fbn})}$$

$$- \frac{Nz^{g2}}{D_c^*} \qquad (A2.46)$$

Substituting from equation (2.90) this may be written as:

$$Y_c SC_r Y_m - \frac{\phi_{h2}}{Y_c^* Y_m^*} = \frac{C_{rn} D_m C_{fbd} - C_{rd} MA_r D_f}{C_{rd} A_e A_q A_r D_f} - \frac{Nz^{g2}}{D_c^*} \qquad (A2.47)$$

Finally, equation (A2.47) may be expressed as:

$$Y_c SC_r Y_m - \frac{\phi_{h2}}{Y_c^* Y_m^*} = T_2^+ + T_2^- \qquad (A2.48)$$

where T_2^+ denotes the first term in equation (A2.47) and T_2^- denotes the second, strictly unstable, term.

(iii) C_{1f} dependent term

From the plant model equations and spectral factor definitions obtain:

$$\frac{\phi_{h1}}{Y_c^* Y_{fd}^*} = \frac{B_p^* A_r B_q^* B_q C_x D_{fd}}{A_q A_\ell A_x D_c^*} \qquad (A2.49)$$

The diophantine equation (2.86) allows the strictly unstable part of equation (A2.49) to be separated as follows:

$$\frac{\phi_{h1}}{Y_c^* Y_{fd}^*} = \frac{X}{A_q A_x A_\ell} + \frac{Zz^{g3}}{D_c^*} \qquad (A2.50)$$

From the system equations and spectral factor definitions obtain:

$$Y_c SC_{1f} Y_{fd} = \frac{D_c C_{fbd} C_{1fn} D_{fd}}{C_{1fd} A_\ell A_q A_r (A_p C_{fbd} + B_p C_{fbn})} \qquad (A2.51)$$

From equations (A2.50) and (A2.51) obtain:

$$Y_c SC_{1f}Y_{fd} - \frac{\phi_{h1}}{Y_c^* Y_{fd}^*} = \frac{D_c C_{fbd} C_{1fn} D_{fd} A_x - XA_r C_{1fd}(A_p C_{fbd} + B_p C_{fbn})}{C_{1fd} A_\ell A_x A_q A_r (A_p C_{fbd} + B_p C_{fbn})}$$

$$- \frac{Zz^{g3}}{D_c^*} \qquad \text{(A2.52)}$$

Substituting from equation (2.90) this may be written as:

$$Y_c SC_{1f}Y_{fd} - \frac{\phi_{h1}}{Y_c^* Y_{fd}^*} = \frac{C_{fbd} C_{1fn} D_{fd} A_x - XA_r C_{1fd} D_f}{C_{1fd} A_\ell A_x A_q A_r D_f} - \frac{Zz^{g3}}{D_c^*} \quad \text{(A2.53)}$$

Finally, equation (A2.53) may be expressed as:

$$Y_c SC_{1f}Y_{fd} - \frac{\phi_{h1}}{Y_c^* Y_{fd}^*} = T_3^+ + T_3^- \qquad \text{(A2.54)}$$

where T_3^+ denotes the first term in equation (A2.53) and T_3^- denotes

the second, strictly unstable, term.

Minimisation

Substituting from equations (A2.42), (A2.48) and (A2.54) into

equation (A2.19), the cost-function integrand may be written:

$$I = (T_1^+ + T_1^-)(T_1^+ + T_1^-)^* + (T_2^+ + T_2^-)(T_2^+ + T_2^-)^* + (T_3^+ + T_3^-)(T_3^+ + T_3^-)^* + \phi_{ol}$$

$$\text{(A2.55)}$$

In equation (A2.55) the T_i^+ terms are stable and the T_i^- terms strictly

unstable for $i = \{1,2,3\}$. In the expansion of equation (A2.55) the

terms $T_i^- T_i^{+*}$ are therefore analytic in $|z| < 1$. In addition, the

terms $T_i^- T_i^{+*}/z$ are also analytic in $|z| < 1$ since the T_i^- terms are

strictly proper in z^{-1}. Thus, using the identity:

$$\oint_c T^- T^{+*} \frac{dz}{z} = - \oint_c T^+ T^{-*} \frac{dz}{z} \qquad \text{(A2.56)}$$

and invoking Cauchy's Theorem, the contour integrals of the cross terms $T_1^+ T_1^{-*}$, $T_1^- T_1^{+*}$ in equation (A2.55) are zero. The cost-function therefore simplifies to:

$$J = \frac{1}{2\pi j} \oint_{|z|=1} \left[\sum_{i=1}^{3} (T_i^+ T_i^{+*} + T_i^- T_i^{-*}) + \phi_{ol} \right] \frac{dz}{z} \qquad (A2.57)$$

Since the terms T_i^- and ϕ_{ol} are independent of the controller the cost-function J is minimised by setting:

$$T_i^+ = 0, \quad i = \{1,2,3\} \qquad (A2.58)$$

(i) Feedback controller

From equations (A2.41) and (A2.42), setting $T_1^+ = 0$ involves:

$$C_{fbn} H - C_{fbd} GA_r = 0 \qquad (A2.59)$$

or:

$$C_{fb} = \frac{GA_r}{H} \qquad (A2.60)$$

(ii) Reference controller

From equations (A2.47) and (A2.48), setting $T_2^+ = 0$ involves:

$$C_{rn} D_m C_{fbd} - C_{rd} MA_r D_f = 0 \qquad (A2.61)$$

or:

$$C_r = \frac{MA_r D_f}{D_m C_{fbd}} \qquad (A2.62)$$

(iii) Feedforward controller

From equations (A2.53) and (A2.54), setting $T_3^+ = 0$ involves:

$$C_{fbd} C_{1fn} D_{fd} A_x - XA_r C_{1fd} D_f = 0 \qquad (A2.63)$$

or:

$$C_{1f} = \frac{XA_r D_f}{D_{fd} A_x C_{fbd}} \qquad (A2.64)$$

Using the definition of C_{1f} in equation (A2.5), the feedforward controller becomes:

$$C_{ff} = \frac{XA_r D_f - C_{fbn} C_x D_{fd}}{D_{fd} A_x C_{fbd}} \qquad (A2.65)$$

Minimum cost

Setting $T_i^+ = 0$, $i = \{1,2,3\}$ in equation (A2.57), the minimum cost is found to be:

$$J_{min} = \frac{1}{2\pi j} \oint_{|z|=1} \left[\sum_{i=1}^{3} (T_i^- T_i^{-*}) + \phi_{ol} \right] \frac{dz}{z} \qquad (A2.66)$$

Stability of T_i^+ terms

Implicit in the above proof is the requirement that the T_i^+ terms are asymptotically stable for $i = \{1,2,3\}$. Stability of the T_i^+ terms may be demonstrated as follows:

(1) T_1^+ term

From equations (A2.41) and (A2.42) obtain:

$$T_1^+ = \frac{C_{fbn} H - C_{fbd} GA_r}{A_q A'_{dx} A_n A_r D_f D_c} \qquad (A2.67)$$

By definition, A_q and A_r are strictly Hurwitz polynomials. By virtue of conditions (a) and (c) in Theorem 6 A'_{dx} is strictly Hurwitz. From Corollary 4 A_n divides both C_{fbn} and G. By definition, D_f is Hurwitz, but in the limiting case when D_f has a zero on the unit circle, this zero will also be in C_{fbn} and G (by virtue of Corollary 6). From Lemma 1 D_c is strictly Hurwitz. T_1^+ is therefore asymptotically stable.

(ii) T_2^+ term

From equations (A2.47) and (A2.48) obtain:

$$T_2^+ = \frac{C_{rn}D_m C_{fbd} - C_{rd}MA_r D_f}{C_{rd}A_e A_q A_r D_f} \qquad (A2.68)$$

Substituting from the implied reference diophantine equation (2.91),
and using equation (A2.62), the expression for T_2^+ may, after some
algebraic manipulation be written as:

$$T_2^+ = \frac{MA'_{pe} (D_m H - C_{rd})}{A'_e A_q D_f D_c D_m} \qquad (A2.69)$$

By definition, A_q is strictly Hurwitz. By virtue of condition (b)
in Theorem 6 A'_e is strictly Hurwitz. By definition, D_f is Hurwitz,
but in the limiting case when D_f has a zero on the unit circle, this
zero will also be in A'_{pe} (by virtue of Corollary 6 and equation
(2.17)). From Lemma 1 D_c and D_m are strictly Hurwitz. T_2^+ is
therefore asymptotically stable.

(iii) T_3^+ term

From equations (A2.53) and (A2.54) obtain:

$$T_3^+ = \frac{C_{fbd}C_{1fn}D_{fd}A_x - XA_r C_{1fd}D_f}{C_{1fd}A_\ell A_x A_r D_f A_q} \qquad (A2.70)$$

Substituting from the implied feedforward diophantine equation
(2.92), and using equation (A2.64), the expression for T_3^+ may, after
some algebraic manipulation, be written as:

$$T_3^+ = \frac{XA'_{p\ell x}(H - C_{fbd})}{A'_{\ell x}A_q D_f D_c} \qquad\qquad (A2.71)$$

By definition, A_q is strictly Hurwitz. By virtue of condition (c) in Theorem 6 $A'_{\ell x}$ is strictly Hurwitz. By definition, D_f is Hurwitz, but in the limiting case when D_f has a zero on the unit circle, this zero will also be in $A'_{p\ell x}$ (by virtue of Corollary 6 and equation (2.27)). From Lemma 1 D_c is strictly Hurwitz. T_3^+ is therefore asymptotically stable.

Solvability conditions

It only remains to relate the conditions (a)-(d) in Theorem 6 to solvability of the optimal control problem. Problem solvability in this context is understood to mean that the given data generate a controller which renders the cost-function finite.

Clearly, the cost will be finite if and only if the twelve transfer-functions in equations (2.39) and (2.40) are asymptotically stable.

Consider the case when α has a zero on the unit circle as discussed in Corollary 6. Such a zero arises when A_p has a zero on the unit circle __and__ when A_d and A_x do not. Using Corollary 6 D_f and C_{fbn} will also have this zero. From equations (A2.62) and (A2.65) C_{rn} and C_{ffn} will also have this zero. As a consequence, this unstable zero will cancel in all transfer-functions in equations (2.39) and (2.40) and the problem remains solvable.

By Corollary 4 (which results from condition (d)) the transfer-functions $B_p C_{fbn} C_n/A_n$ and $C_{fbn} C_n A_p/A_n$ are asymptotically stable. By equation (A2.62) and Lemma 1 the transfer-functions

$B_pC_{rn}C_{fbd}/C_{rd}$ and $C_{rn}A_pC_{fbd}/C_{rd}$ are asymptotically stable. By equation (A2.65), Corollary 5, and Lemma 1 the transfer-functions $B_pC_{ffn}C_{fbd}/C_{ffd}$ and $A_pC_{ffn}C_{fbd}/C_{ffd}$ are asymptotically stable.

Finally, conditions (a) - (c) ensure asymptotic stability of the remaining transfer-functions as follows:

(i)   Condition (a)

Clearly, condition (a) ensures asymptotic stability of the transfer-functions $C_dA_pC_{fbd}/A_d$ and $C_{fbn}C_dA_p/A_d$.

(ii) Condition (b)

The third transfer-function in equation (2.39) is:

$$\frac{(C_{rd}\alpha - B_pC_{rn}C_{fbd})E_r}{A_e\alpha C_{rd}}$$

Substituting from the implied reference diophantine equation (2.91) and using equation (A2.62) this may be rewritten, after some algebraic manipulation, as:

$$\frac{A_qQE_rD_f}{A_e'D_m\alpha}$$

By condition (b) A_e' is strictly Hurwitz and by Lemma 1 so is D_m. This transfer-function is therefore asymptotically stable.

The third transfer-function in equation (2.40) is:

$$\frac{C_{rn}E_rA_pC_{fbd}}{C_{rd}A_e\alpha}$$

This may be rewritten, using equation (A2.62), as:

$$\frac{C_{rn} E_r A_p}{D_m A_e \alpha}$$

Condition (b) and Lemma 1 ensure asymptotic stability of this transfer-function.

(iii) Condition (c)

The fifth transfer-function in equation (2.39) is:

$$\frac{(C_x A_p C_{ffd} - A_x C_{ffn} B_p) E_\ell C_{fbd}}{A_x A_\ell C_{ffd} \alpha}$$

Substituting from the implied feedforward equation (2.92) and using equation (A2.65) this may be rewritten, after some algebraic manipulation, as:

$$\frac{A_q YE_\ell D_f}{A'_{\ell x} D_{fd} \alpha}$$

By condition (c) $A'_{\ell x}$ is strictly Hurwitz and by Lemma 1 so is D_{fd}. This transfer-function is therefore asymptotically stable.

The fifth transfer-function in equation (2.40) is:

$$\frac{(C_{fbn} C_x C_{ffd} + A_x C_{ffn} C_{fbd}) E_\ell A_p}{A_x A_\ell C_{ffd} \alpha}$$

Substituting from equation (A2.65) this may be written:

$$\frac{(C_{fbn} C_x D_{fd} + C_{ffn}) E_\ell A'_{p\ell x}}{A'_{\ell x} D_{fd} \alpha}$$

Again, condition (c) and Lemma 1 ensure asymptotic stability of this transfer-function.

APPENDIX 3 : Proof of Theorem 12

The proof of Theorem 12 follows that of Theorem 1 in Appendix 1 up to equation (A1.21).

Using the common denominator form of the system model given in equations (2.99)-(2.102) and using the polynomial equation form of the cost-function weights given by equation (2.45), the spectral factors (A1.10)-(A1.12) may be written as:

$$Y_c Y_c^* = \frac{BA_r B_q B_q^* A_r^* B^* + AA_q B_r B_r^* A_q^* A^*}{AA_q A_r A_r^* A_q^* A^*} \tag{A3.1}$$

$$Y_f Y_f^* = (A_n C \sigma_d C^* A_n^* + A C_n \sigma_n C_n^* A^* + A_n E \sigma_r E^* A_n^*$$
$$+ AA_n \sigma_{rn} A_n^* A^* + A_n D \sigma_{\ell n} D^* A_n^*)/AA_n A_n^* A^* \tag{A3.2}$$

$$Y_{fd} Y_{fd}^* = \frac{A_\ell \sigma_{\ell n} A_\ell^* + E_\ell \sigma_\ell E_\ell^*}{A_\ell A_\ell^*} \tag{A3.3}$$

Comparison of equations (A1.19)-(A1.21) with equations (A3.1)-(A3.3) then yields:

$$D_c D_c^* = BA_r B_q B_q^* A_r^* B^* + AA_q B_r B_r^* A_q^* A^* \tag{A3.4}$$

$$D_f D_f^* = A_n C \sigma_d C^* A_n^* + A C_n \sigma_n C_n^* A^* + A_n E \sigma_r E^* A_n^*$$
$$+ AA_n \sigma_{rn} A_n^* A^* + A_n D \sigma_{\ell n} D^* A_n^* \tag{A3.5}$$

$$D_{fd} D_{fd}^* = A_\ell \sigma_{\ell n} A_\ell^* + E_\ell \sigma_\ell E_\ell^* \tag{A3.6}$$

and:

$$A_c = AA_q A_r \tag{A3.7}$$

$$A_f = AA_n \tag{A3.8}$$

$$A_{fd} = A_\ell \tag{A3.9}$$

Each of the controller dependent terms in equation (A1.17) may now be simplified separately:

(i) C_c dependent term

From the plant model equations and spectral factor definitions obtain:

$$\frac{\phi_{h2}}{Y_c^* Y_f^*} = \frac{B^* A_r^* B_q^* B_q (D_f D_f^* - AC_n \sigma_n C_n^* A^*)}{AA_q A_n D_c^* D_f^*} \tag{A3.10}$$

The diophantine equation (2.109) allows the strictly unstable part of equation (A3.10) to be separated as follows:

$$\frac{\phi_{h2}}{Y_c^* Y_f^*} = \frac{G}{AA_q A_n} + \frac{Fz^{gl}}{D_c^* D_f^*} \tag{A3.11}$$

From the system equations and spectral factor definitions obtain:

$$Y_c SC_c Y_f = \frac{D_c D_f C_{cn}}{AA_q A_n A_r (AC_{cd} + BC_{cn})} \tag{A3.12}$$

From equations (A3.11) and (A3.12) obtain:

$$Y_c SC_c Y_f - \frac{\phi_{h2}}{Y_c^* Y_f^*} = \frac{D_c D_f C_{cn} - GA_r (AC_{cd} + BC_{cn})}{AA_q A_n A_r (AC_{cd} + BC_{cn})} - \frac{Fz^{gl}}{D_c^* D_f^*} \tag{A3.13}$$

Substituting from the implied cascade diophantine equation (2.117), equation (A3.13) may be expressed as:

$$Y_c SC_c Y_f - \frac{\phi_{h2}}{Y_c^* Y_f^*} = \frac{C_{cn} H - GA_r C_{cd}}{A_q A_n A_r (AC_{cd} + BC_{cn})} - \frac{Fz^{gl}}{D_c^* D_f^*} \tag{A3.14}$$

Finally, equation (A3.14) may be expressed as:

$$Y_c SC_c Y_f - \frac{\phi_{h2}}{Y_c^* Y_f^*} = T_1^+ + T_1^- \tag{A3.15}$$

where T_1^+ denotes the first term in equation (A3.14) and T_2^- denotes the second, strictly unstable, term.

(ii) C_{cf} dependent term

From the plant model equations and spectral factor definitions
obtain:

$$\frac{\phi_{h1}}{Y_c^* Y_{fd}^*} = \frac{B^* A_r^* B_q^* B_q DD_{fd}}{AA_q A_\ell D_c^*} \tag{A3.16}$$

The diophantine equation (2.114) allows the strictly unstable part of
equation (A3.16) to be separated as follows:

$$\frac{\phi_{h1}}{Y_c^* Y_{fd}^*} = \frac{X}{AA_q A_\ell} + \frac{Zz^{g2}}{D_c^*} \tag{A3.17}$$

From the system equations and spectral factor definitions obtain:

$$Y_c SC_{cf} Y_{fd} = \frac{D_c C_{cd} C_{cfn} D_{fd}}{C_{cfd} A_\ell A_q A_r (AC_{cd} + BC_{cn})} \tag{A3.18}$$

From equations (A3.17) and (A3.18) obtain:

$$Y_c SC_{cf} Y_{fd} - \frac{\phi_{h1}}{Y_c^* Y_{fd}^*} = \frac{D_c C_{cd} C_{cfn} D_{fd} A - XA_r C_{cfd}(AC_{cd} + BC_{cn})}{C_{cfd} A_\ell A_q A_r A(AC_{cd} + BC_{cn})}$$
$$- \frac{Zz^{g2}}{D_c^*} \tag{A3.19}$$

Substituting from equations (2.108) and (2.117) this may be written:

$$Y_c SC_{cf} Y_{fd} - \frac{\phi_{h1}}{Y_c^* Y_{fd}^*} = \frac{C_{cd} C_{cfn} D_{fd} A - XA_r C_{cfd} D_f}{C_{cfd} A_\ell A_q A_r AD_f} - \frac{Zz^{g2}}{D_c^*} \tag{A3.20}$$

Finally, equation (A3.20) may be expressed as:

$$Y_c SC_{cf} Y_{fd} - \frac{\phi_{h1}}{Y_c^* Y_{fd}^*} = T_2^+ + T_2^- \tag{A3.21}$$

where T_2^+ denotes the first term in equation (A3.20) and T_2^- denotes
the second, strictly unstable, term.

Minimisation

Using a similar argument to the one used in Appendix 1 the cost function may be minimised by setting $T_i^+ = 0$, $i = \{1,2\}$.

(i) Cascade controller

From equations (A3.14) and (A3.15), setting $T_1^+ = 0$ involves:

$$C_{cn}H - GA_rC_{cd} = 0 \tag{A3.22}$$

or:

$$C_c = \frac{GA_r}{H} \tag{A3.23}$$

(ii) Feedforward controller

From equations (A3.20) and (A3.21), setting $T_2^+ = 0$ involves:

$$C_{cd}C_{cfn}D_{fd}A - XA_rC_{cfd}D_f = 0 \tag{A3.24}$$

or:

$$C_{cf} = \frac{XA_rD_f}{C_{cd}D_{fd}A} \tag{A3.25}$$

Using the definition of C_{cf} in equation (A1.5), the feedforward controller becomes:

$$C_{ff} = \frac{XA_rD_f - C_{cn}DD_{fd}}{D_{fd}AC_{cd}} \tag{A3.26}$$

Solvability conditions

To verify the solvability conditions (a)-(c) in Theorem 12 for the optimal control problem using a common denominator model it is sufficient to show that:

    (i)   The conditions (a)-(c) in Theorem 12 are equivalent to conditions (a)-(d) in Theorem 1.

(ii)   The controllers generated by the equations in Theorem 12
and in Theorem 1 are the same.

(i) Equivalence of solvability conditions

Any factors of A_d, A_e or A_x which are not also factors of A_p
will result in common factors in A and B. Condition (a) in Theorem
12 therefore subsumes conditions (a) and (b), and the A_x part of
condition (c), in Theorem 1.

Condition (b) in Theorem 12, that any unstable factor of A_ℓ must
be a factor of A and D means firstly that any unstable factors of A_ℓ
must be in A_p, and secondly that the product of such factors with any
which are also in A_x must appear in A_p (i.e. this is just condition
(c) in Theorem 1). Condition (c) in Theorem 12 is clearly equivalent
to condition (d) in Theorem 1.

(ii) Equivalence of controllers

From the definitions in Table 2.2:

$$A = A_{dex} A'_{pf} = A_p A'_{dexp} \tag{A3.27}$$

$$B = B_p A'_{dexp} \tag{A3.28}$$

$$C = A'_{ex} C_d A'_{pf} \tag{A3.29}$$

$$D = A'_{ed} C_x A'_{pf} \tag{A3.30}$$

$$E = A'_{xd} E_r A'_{pf} \tag{A3.31}$$

Using equations (A3.27) - (A3.31) equations (2.50) and (2.51)
become:

$$D_c^* D_f^* z^{-g_1} G + FAA_q A_n = B_p^* A_r^* B_q^* B_q^* R_1 \tag{A3.32}$$

$$D_c^* D_f^* z^{-g_1} H - FBA_r A_q A_n = A_p^* R_2 \tag{A3.33}$$

where:

$$R_1 = z^{-g1}(D_f D_f^* - C_n \sigma_n C_n^* AA^*) \tag{A3.34}$$

$$R_2 = z^{-g1}(D_f D_f^* A_q A_q^* B_r B_r^* + BB^* A_r A_r^* B_q B_q^* C_n \sigma_n C_n^*) \tag{A3.35}$$

Multiplying equations (A3.32) and (A3.33) by A'^*_{dexp} obtain:

$$D_c'^* D_f^* z^{-g1} G + F'AA_q A_n = B^* A_r^* B_q^* B_q R_1 \tag{A3.36}$$

$$D_c'^* D_f^* z^{-g1} H - F'BA_r A_q A_n = A^* R_2 \tag{A3.37}$$

where:

$$D_c' = D_c A'_{dexp} \tag{A3.38}$$

$$F' = FA'^*_{dexp} \tag{A3.39}$$

From (A3.38) and (2.46) obtain:

$$D_c'D_c'^* = BA_r B_q B_q^* A_r^* B^* + AA_q B_r B_r^* A_q^* A^* \tag{A3.40}$$

Using equations (A3.27)-(A3.31) it is clear that the definitions of D_f in equations (2.47) and (2.106) are equivalent.

Comparison of equations (A3.36), (A3.37) and (A3.40) with equations (2.109), (2.110) and (2.105) shows the solution G,H of equations (A3.36) and (A3.37) with:

$$(D_c'^* z^{-g1})^{-1} F' \quad \text{strictly proper} \tag{A3.41}$$

to be equivalent to the solution G,H of equations (2.109) and (2.110) with $(D_c^* z^{-g1})^{-1} F$ strictly proper.

From the definitions in Table 2.2:

$$A = A_x A'_{ed} A'_{pf} \tag{A3.42}$$

Multiplying equations (2.55) and (2.56) by $A'_{ed} A'_{pf}$ and using equations (A3.30) and (A3.42) obtain:

$$D_c^* z^{-g2} X' + ZAA_q A_\ell = z^{-g2} B_p^* A_r^* B_q^* B_q DD_{fd} \tag{A3.43}$$

$$D_c^* z^{-g2} Y' - ZB_p A_r A'_{\ell x} A'_{ed} A'_{pf} = z^{-g2} A_p^* A_q^* B_r^* B_r A'_{p\ell x} DD_{fd} \tag{A3.44}$$

where:

$$X' = XA'_{ed}A'_{pf} \quad , \quad Y' = YA'_{ed}A'_{pf} \tag{A3.45}$$

Using the definitions in Table 2.2 obtain:

$$B_p A'_{\ell x} A'_{ed} A'_{pf} = \frac{B_p A_\ell A_x A'_{ed} A'_{pf}}{D_{p\ell x}}$$

$$= B_p A'_{dexp} \frac{A_p}{D_{p\ell x}} A_\ell$$

$$= BA'_{p\ell x} A_\ell$$

Substituting in equation (A3.44) obtain:

$$D_c^* z^{-g2} Y'' - ZBA_r A_\ell = z^{-g2} A_p^* A_q^* B_r^* B_r DD_{fd} \tag{A3.46}$$

where $Y' \triangleq Y'' A'_{p\ell x}$. Multiplying equations (A3.43) and (A3.46) by A'^*_{dexp} obtain:

$$D_c'^* z^{-g2} X' + Z' AA_q A_\ell = z^{-g2} B^* A_r^* B_q^* B_q DD_{fd} \tag{A3.47}$$

$$D_c'^* z^{-g2} Y'' - Z' BA_r A_\ell = z^{-g2} A^* A_q^* B_r^* B_r DD_{fd} \tag{A3.48}$$

where:

$$Z' = ZA'^*_{dexp} \tag{A3.49}$$

Comparison of equations (A3.47) and (A3.48) with equations (2.114) and (2.115) shows the solution X', Y" of equations (A3.47) and (A3.48) with:

$$(D_c'^* z^{-g2})^{-1} Z' \quad \text{strictly proper}$$

to be equivalent to the solution X,Y of equations (2.114) and (2.115) with $(D_c^* z^{-g2})^{-1} Z$ strictly proper. The controllers obtained from equations (2.54) and (2.113) are therefore equivalent.

APPENDIX 4 : Proof of Theorem 14

The proof of Theorem 14 follows that of Theorem 6 in Appendix 2 up to equation (A2.24).

Using the common denominator form of the system model given in equations (2.125)–(2.127), and using the polynomial equation form of the cost-function weights given by equation (2.45), the spectral factors (A2.10)–(A2.13) may be written as:

$$Y_c Y_c^* = \frac{BA_r B_q B A_q^* A_r^* B^* + AA_q B_r B A_r^* A_q^*}{AA_q A_r A_r^* A_q^*} \tag{A4.1}$$

$$Y_f Y_f^* = \frac{A_n C\sigma_d C^* A_n^* + AC_n \sigma_n C_n^* A^* + A_n D\sigma_{\ell n} D^* A_n^*}{AA_n A_n^*} \tag{A4.2}$$

$$Y_{fd} Y_{fd}^* = \frac{A_\ell \sigma_{\ell n} A_\ell^* + E_\ell \sigma_\ell E_\ell^*}{A_\ell A_\ell^*} \tag{A4.3}$$

$$Y_m Y_m^* = \frac{A_e \sigma_{rn} A_e^* + E_r \sigma_r E_r^*}{A_e A_e^*} \tag{A4.4}$$

Comparison of equations (A2.21)–(A2.24) with equations (A4.1)–(A4.4) yields:

$$D_c D_c^* = BA_r B_q B_q^* A_r^* B^* + AA_q B_r B A_r^* A_q^* A^* \tag{A4.5}$$

$$D_f D_f^* = A_n C\sigma_d C^* A_n^* + AC_n \sigma_n C_n^* A^* + A_n D\sigma_{\ell n} D^* A_n^* \tag{A4.6}$$

$$D_{fd} D_{fd}^* = A_\ell \sigma_{\ell n} A_\ell^* + E_\ell \sigma_\ell E_\ell^* \tag{A4.7}$$

$$D_m D_m^* = A_e \sigma_{rn} A_e^* + E_r \sigma_r E_r^* \tag{A4.8}$$

and:

$$A_c = AA_q A_r \tag{A4.9}$$

$$A_f = AA_n \tag{A4.10}$$

$$A_{fd} = A_\ell \tag{A4.11}$$

$$A_m = A_e \tag{A4.12}$$

Each of the controller dependent terms in equation (A2.19) may now be simplified separately:

(i) $\underline{C_{fb} \text{ dependent term}}$

From the plant model equations and spectral factor definitions obtain:

$$\frac{\phi_{h3}}{Y_c^* Y_f^*} = \frac{B^* A_r B_q^* B_q (D_f D_f^* - AC_n \sigma_n C_n^* A^*)}{AA_q A_n D_c^* D_f^*} \tag{A4.13}$$

The diophantine equation (2.137) allows the strictly unstable part of equation (A4.13) to be separated as follows:

$$\frac{\phi_{h3}}{Y_c^* Y_f^*} = \frac{G}{AA_q A_n} + \frac{Fz^{g1}}{D_c^* D_f^*} \tag{A4.14}$$

From the system equations and spectral factor definitions obtain:

$$Y_c MY_f = \frac{D_c D_f C_{fbn}}{AA_q A_r A_n (AC_{fbd} + BC_{fbn})} \tag{A4.15}$$

From equations (A4.14) and (A4.15) obtain:

$$Y_c MY_f - \frac{\phi_{h3}}{Y_c^* Y_f^*} = \frac{D_c D_f C_{fbn} - GA_r (AC_{fbd} + BC_{fbn})}{AA_q A_r A_n (AC_{fbd} + BC_{fbn})}$$
$$- \frac{Fz^{g1}}{D_c^* D_f^*} \tag{A4.16}$$

Substituting from the implied feedback diophantine equation (2.148), equation (A4.16) may be expressed as:

$$Y_c MY_f - \frac{\phi_{h3}}{Y_c^* Y_f^*} = \frac{C_{fbn} H - GA_r C_{fbd}}{A_q A_r A_n (AC_{fbd} + BC_{fbn})}$$
$$- \frac{Fz^{g1}}{D_c^* D_f^*} \tag{A4.17}$$

Finally, equation (A4.17) may be expressed as:

$$Y_c MY_f - \frac{\phi_{h3}}{Y_c^* Y_f^*} = T_1^+ + T_1^-$$ (A4.18)

where T_1^+ denotes the first term in equation (A4.17) and T_1^- denotes the second, strictly unstable, term.

(ii) C_r dependent term

From the plant model equations and spectral factor definitions obtain:

$$\frac{\phi_{h2}}{Y_c^* Y_m^*} = \frac{B^* A_r B_q^* B_q D_m}{A_q A_e D_c^*}$$ (A4.19)

The diophantine equation (2.142) allows the strictly unstable part of equation (A4.19) to be separated as follows:

$$\frac{\phi_{h2}}{Y_c^* Y_m^*} = \frac{M}{A_q A_e} + \frac{Nz^{g2}}{D_c^*}$$ (A4.20)

From the system equations and spectral factor definitions obtain:

$$Y_c SC_r Y_m = \frac{D_c C_{fbd} D_m C_{rn}}{C_{rd} A_e A_q A_r (AC_{fbd} + BC_{fbn})}$$ (A4.21)

From equations (A4.20) and (A4.21) obtain:

$$Y_c SC_r Y_m - \frac{\phi_{h2}}{Y_c^* Y_m^*} = \frac{D_c C_{rn} D_m C_{fbd} - C_{rd} MA_r (AC_{fbd} + BC_{fbn})}{C_{rd} A_e A_q A_r (AC_{fbd} + BC_{fbn})}$$

$$- \frac{Nz^{g2}}{D_c^*}$$ (A4.22)

Substituting from equations (2.136) and (2.148) this may now be written:

$$Y_c SC_r Y_m - \frac{\phi_{h2}}{Y_c^* Y_m^*} = \frac{C_{rn} D_m C_{fbd} - C_{rd} MA_r D_f}{C_{rd} A_e A_q A_r D_f} - \frac{Nz^{g2}}{D_c^*} \tag{A4.23}$$

Finally, equation (A4.23) may be expressed as:

$$Y_c SC_r Y_m - \frac{\phi_{h2}}{Y_c^* Y_m^*} = T_2^+ + T_2^- \tag{A4.24}$$

where T_2^+ denotes the first term in equation (A4.23) and T_2^- denotes the second, strictly unstable, term.

(iii) $\underline{C_{1f} \text{ dependent term}}$

From the plant model equations and spectral factor definitions obtain:

$$\frac{\phi_{h1}}{Y_c^* Y_{fd}^*} = \frac{B^* A_r^* B_q^* B_q DD_{fd}}{AA_q A_\ell D_c^*} \tag{A4.25}$$

The diophantine equation (2.145) allows the strictly unstable part of equation (A4.25) to be separated as follows:

$$\frac{\phi_{h1}}{Y_c^* Y_{fd}^*} = \frac{X}{AA_q A_\ell} + \frac{Zz^{g3}}{D_c^*} \tag{A4.26}$$

From the system equations and spectral factor definitions obtain:

$$Y_c SC_{1f} Y_{fd} = \frac{D_c C_{fbd} C_{1fn} D_{fd}}{C_{1fd} A_\ell A_q A_r (AC_{fbd} + BC_{fbn})} \tag{A4.27}$$

From equations (A4.26) and (A4.27) obtain:

$$Y_c SC_{1f} Y_{fd} - \frac{\phi_{h1}}{Y_c^* Y_{fd}^*} = \frac{D_c C_{fbd} C_{1fn} D_{fd} A - XA_r C_{1fd}(AC_{fbd} + BC_{fbn})}{C_{1fd} AA_q A_\ell A_r (AC_{fbd} + BC_{fbn})}$$
$$- \frac{Zz^{g3}}{D_c^*} \tag{A4.28}$$

Substituting from equations (2.136) and (2.148) this may now be written:

$$Y_c S C_{1f} Y_{fd} - \frac{\phi_{h1}}{Y_c^* Y_{fd}^*} = \frac{C_{fbd} C_{1fn} D_{fd} A - X A_r C_{1fd} D_f}{C_{1fd} A A_q A_\ell A_r D_f}$$
$$- \frac{Z z^{g3}}{D_c^*} \tag{A4.29}$$

Finally, equation (A4.29) may be expressed as:

$$Y_c S C_{1f} Y_{fd} - \frac{\phi_{h1}}{Y_c^* Y_{fd}^*} = T_3^+ + T_3^- \tag{A4.30}$$

where T_3^+ denotes the first term in equation (A4.29) and T_3^- denotes the second, strictly unstable, term.

Minimisation

Using a similar argument to the one used in Appendix 2 the cost-function may be minimised by setting $T_i^+ = 0$, $i = \{1,2,3\}$.

(i) Feedback controller

From equations (A4.17) and (A4.18), setting $T_1^+ = 0$ involves:

$$C_{fbn} H - G A_r C_{fbd} = 0 \tag{A4.31}$$

or:

$$C_{fb} = \frac{G A_r}{H} \tag{A4.32}$$

(ii) Reference controller

From equations (A4.23) and (A4.24), setting $T_2^+ = 0$ involves:

$$C_{rn} D_m C_{fbd} - C_{rd} M A_r D_f = 0 \tag{A4.33}$$

or:

$$C_r = \frac{MA_r D_f}{D_m C_{fbd}} \qquad (A4.34)$$

(iii) <u>Feedforward controller</u>

From equations (A4.29) and (A4.30), setting $T_3^+ = 0$ involves:

$$C_{fbd} C_{1fn} D_{fd} A - XA_r C_{1fd} D_f = 0 \qquad (A4.35)$$

or:

$$C_{1f} = \frac{XA_r D_f}{D_{fd} AC_{fbd}} \qquad (A4.36)$$

Using the definition of C_{1f} in equation (A2.5), the feedforward controller becomes:

$$C_{ff} = \frac{XA_r D_f - C_{fbn} DD_{fd}}{D_{fd} AC_{fbd}} \qquad (A4.37)$$

<u>Solvability conditions</u>

To verify the solvability conditions (a) - (d) in Theorem 14 it is sufficient to show that:

(i) The conditions (a) - (d) in Theorem 14 are equivalent to conditions (a) - (d) in Theorem 6.

(ii) The controllers generated by the equations in Theorem 14 and in Theorem 6 are the same.

(i) <u>Equivalence of solvability conditions</u>

Any factors of A_d or A_x which are not also factors of A_p will appear as common factors in A and B. Condition (a) in Theorem 14 therefore subsumes condition (a), and the A_x part of condition (c), in Theorem 6.

Condition (c) in Theorem 14 means that any unstable factors of A_ℓ must be in A_p, and that the product of such factors with any which are also in A_x must appear in A_p (i.e. this is just condition (c) in Theorem 6).

Conditions (b) and (d) in Theorem 14 are clearly equivalent to conditions (b) and (d), respectively, in Theorem 6.

(ii) Equivalence of controllers

From the definitions in Table 2.2:

$$A = A_{dx}A_p' = A_pA_{dx}' \qquad (A4.38)$$

$$B = B_pA_{dx}' \qquad (A4.39)$$

$$C = A_x'C_dA_p' \qquad (A4.40)$$

$$D = A_d'C_xA_p' \qquad (A4.41)$$

Using equations (A4.38) - (A4.41) equations (2.78) and (2.79) become:

$$D_c^*D_f^*z^{-gl}G + FAA_qA_n = B_p^*A_r^*B_q^*B_qR_1 \qquad (A4.42)$$

$$D_c^*D_f^*z^{-gl}H - FBA_rA_qA_n = A_p^*R_2 \qquad (A4.43)$$

where:

$$R_1 = z^{-gl}(D_fD_f^* - C_n\sigma_nC_n^*AA^*) \qquad (A4.44)$$

$$R_2 = z^{-gl}(D_fD_f^*A_qA_q^*B_rB_r^* + BB^*A_rA_r^*B_qB_q^*C_n\sigma_nC_n^*) \qquad (A4.45)$$

Multiplying equations (A4.42) and (A4.43) by $A_{dx}'^*$ obtain:

$$D_c'^*D_f^*z^{-gl}G + F'AA_qA_n = B^*A_r^*B_q^*B_qR_1 \qquad (A4.46)$$

$$D_c'^*D_f^*z^{-gl}H - F'BA_rA_qA_n = A^*R_2 \qquad (A4.47)$$

where:

$$D'_c = D_c A'_{dx} \tag{A4.48}$$

$$F' = F A'^*_{dx} \tag{A4.49}$$

From (A4.48) and (2.73) obtain:

$$D'_c D'^*_c = BA_r B_q B^*_q A^*_r B^* + AA_q B_r B^*_r A^*_q A^* \tag{A4.50}$$

Using equations (A4.38) – (A4.41) it is clear that the definitions of D_f in equations (2.74) and (2.133) are equivalent.

Comparison of equations (A4.46), (A4.47) and (A4.50) with equations (2.137), (2.138) and (2.132) shows the solution G,H of equations (A4.46) and (A4.47) with:

$$(D'^*_c z^{-g1})^{-1} F' \quad \text{strictly proper}$$

to be equivalent to the solution G,H of equations (2.137) and (2.138) with $(D^*_c z^{-g1})^{-1} F$ strictly proper.

Denote by A''_{dx} that part of A'_{dx} which does not have any common factors with A_e. Then:

$$A''_{dx} A'_e = A'_{dx} A'_{ec} \tag{A4.51}$$

$$A''_{dx} A'_{pe} = A' \tag{A4.52}$$

Multiplying equation (2.84) by A''_{dx} obtain:

$$D^*_c z^{-g2} Q' - NBA_r A'_{ec} = z^{-g2} A^*_p A^*_q B^*_r A' D_m \tag{A4.53}$$

where:

$$Q' = QA''_{dx} \tag{A4.54}$$

Multiplying equations (2.83) and (A4.53) by A'^*_{dx} obtain:

$$D'^*_c z^{-g2} M + N' A_q A_e = z^{-g2} B^* A^*_r B^*_q B_q D_m \tag{A4.55}$$

$$D'^*_c z^{-g2} Q' - N' BA_r A'_{ec} = z^{-g2} A^* A^*_q B_r A' D_m \tag{A4.56}$$

where:

$$N' = NA'^{*}_{dx} \tag{A4.57}$$

Comparison of equations (A4.55) and (A4.56) with equations (2.142) and (2.143) shows the solution M,Q' of equations (A4.55) and (A4.56) with:

$$(D'^{*}_{c}z^{-g2})^{-1}N' \quad \text{strictly proper}$$

to be equivalent to the solution M,Q of equations (2.142) and (2.143) with $(D^{*}_{c}z^{-g2})^{-1}N$ strictly proper.

From the definitions in Table 2.2:

$$A = A_x A'_d A'_p \tag{A4.58}$$

Multiplying equations (2.86) and (2.87) by $A'_d A'_p$ and using equations (A4.41) and (A4.58) obtain:

$$D^{*}_{c}z^{-g3}X' + ZAA_q A_\ell = z^{-g3}B_p A^{*}_r B^{*}_q B_q DD_{fd} \tag{A4.59}$$

$$D^{*}_{c}z^{-g3}Y' - ZB_p A_r A'_{\ell x} A'_d A'_p = z^{-g3}A_p A^{*}_q A^{*}_r B^{*}_r B A'_{p\ell x} DD_{fd} \tag{A4.60}$$

where:

$$X' = XA'_d A'_p \quad , \quad Y' = YA'_d A'_p \tag{A4.61}$$

Using the definitions in Table 2.2 obtain:

$$
\begin{aligned}
B_p A'_{\ell x} A'_d A'_p &= \frac{B_p A_\ell A_x A'_d A'_p}{D_{p\ell x}} \\
&= B_p A'_{dx} \frac{A_p}{D_{p\ell x}} A_\ell \\
&= BA'_{p\ell x} A_\ell
\end{aligned}
$$

Substituting in equation (A4.60) obtain:

$$D^{*}_{c}z^{-g3}Y'' - ZBA_r A_\ell = z^{-g3}A^{*}_p A^{*}_q A^{*}_r B^{*}_r DD_{fd} \tag{A4.62}$$

where $Y' = Y''A'_{p\ell x}$. Multiplying equations (A4.59) and (A4.62) by A'^{*}_{dx} obtain:

$$D_c'^{*}z^{-g3}X' + Z'AA_qA_\ell = z^{-g3}B^{*}A_r^{*}B_q^{*}B_q DD_{fd} \qquad (A4.63)$$

$$D_c'^{*}z^{-g3}Y'' - Z'BA_rA_\ell = z^{-g3}A^{*}A_q^{*}B_r^{*}B_r DD_{fd} \qquad (A4.64)$$

where:

$$Z' = ZA_{dx}'^{*} \qquad (A4.65)$$

Comparison of equations (A4.63) and (A4.64) with equations (2.145)

and (2.146) shows the solution X', Y" of equations (A4.63) and

(A4.64) with:

$$(D_c'^{*}z^{-g3})^{-1}Z' \quad \text{strictly proper} \qquad (A4.66)$$

to be equivalent to the solution X,Y of equations (2.145) and (2.146)

with $(D_c^{*}z^{-g3})^{-1}Z$ strictly proper. The controllers obtained from

equations (2.85) and (2.144) are therefore equivalent.

Lecture Notes in Control and Information Sciences

Edited by M. Thoma and A. Wyner

Lecture Notes in Control and Information Sciences

Edited by M. Thoma and A. Wyner

Lecture Notes in Control and Information Sciences

Edited by M. Thoma and A. Wyner